SPEEDWELL
SCHOOL.

69/1.   68/66.

EXPERIMENTAL PHYSICAL CHEMISTRY

Experimental
Physical Chemistry

by

D. G. DAVIES, B.Sc., A.R.I.C.
Senior Chemistry Master, Palmer's School

and

T. V. G. KELLY, B.Sc.
Assistant Chemistry Master, Palmer's School

MILLS & BOON LIMITED
50 GRAFTON WAY, FITZROY SQUARE
LONDON, W.1

Made and printed in Great Britain by
William Clowes and Sons, Limited, London and Beccles

Preface

The aim of this book is to provide a source of well-tested experiments for the use of students working for G.C.E. Advanced level, University Open Awards, H.N.C. and L.R.I.C. examinations.

Experimental work in physical chemistry has been much neglected at the pre-university level. We suspect there is still an impression that laboratory work in this 'branch' of the subject requires expensive apparatus, a great deal of time and the mastery of difficult techniques if successful results are to be obtained. It has been our experience, however, that, providing experiments are chosen carefully and students are given accurate and detailed practical scripts to work from, even the less able experimenter can achieve satisfying results during normal practical sessions and with relatively simple apparatus.

The overriding consideration in the choice of experiments for this book has been that an experiment must work well when performed by students of a wide range of ability under normal class conditions. 'Working well' includes the requirement that quantitative results should be consistently good. Several of the 'classical' experiments described in theoretical (and some practical) works have been rejected because they do not fulfil this condition.

The text for each of Experiments 1 to 40 includes:

(i) a detailed list of apparatus and chemicals, which enables requirements to be assembled quickly by either experimenter or laboratory steward, so that work may begin without delay;
(ii) detailed instructions for procedure;
(iii) assistance with the recording and analysis of results, graphical analysis being encouraged wherever possible;
(iv) theoretical background, where appropriate, (an attempt having been made to write this in such a way that it provides a link between the pure theory of the lecture room and the practice of the laboratory).

We are aware that a possible criticism is that too much help is given to the abler student. However, a class book of this type must provide for students of varying degrees of ability, and we believe there is always room for the abler student to use his own initiative in the development of his experimental technique, and to give more extensive and detailed consideration to the analysis of results. Experiments 41 to 50 have been given in outline only; their purpose is to give the abler student experience in the planning of an experiment and to encourage him to develop further the practical and deductive skills acquired by performing a selection of the earlier experiments.

An elementary treatment of errors is included in the appendix for those students whose tutors wish to deal with this important facet of experimental work at the pre-university level.

The terms 'litre' and 'millilitre' have been retained as special names for the units dm^3 and cm^3, and it is in no way implied that they are defined units in themselves.

We wish to thank the many sixth-form students at Palmer's for their invaluable assistance in the laboratory. Their comments have resulted in many improvements in this published form of our work. Our special thanks are due to Dr. D. J. Waddington, Dr. L. C. Roselaar and Mr. P. C. Drewry, all of whom read the manuscript and made valuable suggestions for improvement.

Palmer's
1967

D.G.D.
T.K.

REAGENTS AND STANDARD SOLUTIONS REQUIRED FOR THE EXPERIMENTS IN THIS BOOK

A comprehensive list is given in Appendix (vi), p. 124. All of the reagents, with the possible exception of two ion exchange resins, will be readily available in laboratories used for work at and beyond G.C.E. Advanced level. Every effort has been made to keep their number and cost as low as possible.

The authors strongly recommend that stocks of most of the solutions are kept in the laboratory. Large volumes are rarely used and 2½-litre (Winchester) stocks are quite adequate. The purchase of standard (N/1) volumetric solutions, particularly of the common acids and alkalis, pays heavy dividends in time-saving and accuracy of results; the accurate, yet simple, dilution of these to give other required standards is highly economical and eliminates the tedious preliminary of preparing standard solutions from solid or concentrated reagents.

Apparatus

The majority of experiments described in this book can be performed with inexpensive apparatus already available in laboratories used for work at Advanced, Open Scholarship and Higher National Certificate levels.

A thermostatically controlled water bath is desirable but not essential, and most of the experiments in which the use of one is recommended still give very good results if one is not available.

One piece of apparatus which may not be widely used is a simple student polarimeter (Experiments 34 and 35). At least two firms* supply these robust instruments at moderately low cost, and the authors have been highly impressed by the performance of the one in their possession.

---

* W. B. Nicholson Ltd., Thornliebank, Glasgow, and Griffin and George Ltd., Alperton, Wembley, Middlesex.

NOTES FOR THE EXPERIMENTER

1. Read carefully through the text of the experiment. Make sure that you understand both the theoretical background and the experimental techniques involved before you start work.

2. Collect all of the apparatus, solutions and reagents required for the experiment. If everything has already been put out for you, check to see that nothing is missing. A small but vital piece of apparatus missing at a critical moment can ruin the results of an experiment.

3. Assemble the apparatus carefully, making sure that all connections and bungs are secure. If an electrical circuit is being used be especially careful, since a wrong connection can easily result in damage to expensive galvanometers.

4. Record all your experimental work in a notebook kept exclusively for the purpose. An account should include:

(a) THE TITLE or purpose of the experiment.

(b) A DIAGRAM of the apparatus, where appropriate.

(c) A BRIEF RECORD of the experimental operations—there is no virtue in just copying the experimental instructions given, such detail is not required.

(d) A TABLE AND (where applicable) GRAPH OF RESULTS. **All** readings should be recorded in your notebook (never on scrap paper) **at the time they are made**. If the results are to be analysed graphically upon completion of the experiment it is advisable to plot a rough curve, wherever possible, **as the experiment proceeds**. In this way the course of the experiment can be followed and any major deviations checked at the time.

(e) A FULLY EXPLAINED CALCULATION of the final result(s). A final result should be expressed only to that degree of accuracy merited by the method and instruments used. The probable error in the result should be clearly indicated (see Appendix (vii), p. 125). Units must always be given, together with the conditions (*e.g.* room or thermostat temperature) under which an experimental result was obtained.

(f) A SUMMARY OF THE THEORETICAL BACKGROUND where the experimenter feels this might be helpful for future reference.

CONTENTS

ix

Appendix

Solubility Determinations

EXPERIMENT 1. TO CONSTRUCT A SOLUBILITY CURVE FOR POTASSIUM NITRATE IN WATER BY A COOLING METHOD

APPARATUS AND CHEMICALS REQUIRED

400-ml tall beaker; tripod; bunsen; boiling tube; 0–110°C (in 1°C) thermometer; burette; two burette stands and clamps; ring stirrer.

Potassium nitrate (15 g).

EXPERIMENTAL PROCEDURE

Three-quarters fill the 400-ml beaker with water and bring nearly to the boil.

Weigh 15 g of potassium nitrate crystals (to nearest 0·1 g) into a clean dry boiling tube. Charge a burette with distilled water and adjust to the '0' mark. Run 10 ml of water into the tube containing the salt, insert a ring stirrer and thermometer and immerse in the hot water (at about 80°C). Stir until all the solid has dissolved.

Remove the tube from the hot water bath, dry and clamp on a burette stand away from draughts. Stir continuously and note the temperature when distinct crystals first appear (a black background or a light shone from behind facilitates this observation).

Add a further 2·5 ml of distilled water and repeat. Do likewise after further additions of 2·5, 5·0, 5·0 and 10·0 ml of distilled water.

Tabulate the results and plot concentration (g of solid/100 g of water) against crystallisation temperature.

RESULTS

Initial weight of potassium nitrate: 15 g.

Volumes of water added (ml)	Total volume of water (ml)	Concentration (g KNO$_3$ per 100 g water)	Crystallisation temperature (°C)
10·0	10·0		
2·5	12·5		
2·5	15·0		
5·0	20·0		
5·0	25·0		
10·0	35·0		

EXPERIMENT 2. TO DETERMINE THE SOLUBILITY OF AMMONIA IN WATER

APPARATUS AND CHEMICALS REQUIRED

Boiling tube with side arm, and fitted with rubber bung, tubes and clips as in Fig. 2.1; three 250-ml conical flasks; two 0–110°C (in 1°C) thermometers; disused 25-ml pipette with tip removed; 25-ml pipette; 50-ml measuring cylinder; 2-litre graduated flask; burette and stand; 25°C thermostat or a litre beaker; squeeze type wash bottle; white tile; two retort stands and clamps; tripod; bunsen.

0·880 ammonia (75 ml); 0·2M (0·2N) hydrochloric acid (50 ml); methyl red indicator.

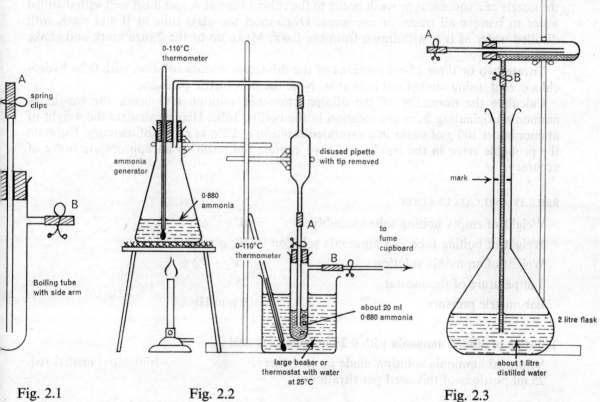

Fig. 2.1   Fig. 2.2   Fig. 2.3

EXPERIMENTAL PROCEDURE

Prepare the assembly shown in Fig. 2.1. The lengths of rubber tubing should be between 3 cm and 5 cm. Weigh the assembly, complete with clips, to the nearest 0·01 g.

Assemble the apparatus shown in Fig. 2.2 with approximately 50 ml of 0·880 ammonia in the ammonia generator and about 20 ml in the weighed boiling tube. If a thermostat at

25°C is not available the water in the litre beaker must be kept at about 25°C by the occasional addition of a little hot water. A tube leading to an efficient fume cupboard must be attached at B.

Remove clips A and B and **gently** heat the ammonia generator to give a slow bubble rate of gas. Continue to generate ammonia for 20–30 min, keeping the gas generator temperature below 60°C. Remove the heat supply and immediately disconnect the pipette at A and fume cupboard lead at B. Firmly replace the clips, and then dry the outside of the assembly and reweigh.

Place about a litre of distilled water in the 2-litre graduated flask. Fill a length of glass tubing with distilled water (a finger closing one end) and attach the open end to the boiling tube assembly at B. With the boiling tube horizontal, lower the glass tube into the 2-litre flask so that it dips well into the distilled water as in Fig. 2.3. Clamp the boiling tube. Open clip B and then clip A and allow the ammonia solution to drain into the flask. Attach the nozzle of a squeeze type wash bottle to the rubber tube at A and flush well with distilled water to remove all traces of ammonia. Disconnect the glass tube at B and wash with distilled water as it is withdrawn from the flask. Make up to the 2-litre mark and shake well.

Titrate two or three 25-ml portions of the diluted ammonia solution with 0·2N hydrochloric acid, using methyl red indicator. Note the barometric pressure.

Calculate the normality of the diluted ammonia solution and hence the weight of ammonia originating from the solution in the boiling tube. Hence calculate the weight of ammonia per 100 g of water in a saturated solution at 25°C at P mm of mercury. Estimate the probable error in the result, and then express the result to an appropriate order of accuracy.

RESULTS AND CALCULATION

Weight of empty boiling tube assembly	$= W$ g
Weight of boiling tube and ammonia solution	$= W_1$ g
Weight of ammonia solution	$= (W_1 - W)$ g
Temperature of thermostat	$= 25°C$
Barometric pressure	$= P$ mm Hg

Titration of diluted ammonia with 0·2N hydrochloric acid

Saturated ammonia solution made up to 2 litres;   Indicator: methyl red.
25 ml portions of this used per titration.

Burette (finish)			
Burette (start)			
Titre (ml)			

25 ml diluted ammonia solution $\equiv$ V ml 0·2N hydrochloric acid

$$\therefore \text{ Normality of diluted ammonia} = \frac{0 \cdot 2 \times V}{25} \text{ N}$$

$\therefore$ Weight of ammonia gas in the 2 litres of diluted solution

$$= \frac{0 \cdot 2 \times V \times 2 \times 17}{25} \text{ g} \quad (\text{E.W. NH}_3 = 17)$$

$$= W_2 \text{ g}$$

= weight of ammonia in the original $(W_1 - W)$ g of saturated solution

$\therefore$ Weight of water in $(W_1 - W)$ g saturated solution $= (W_1 - W) - W_2$ g

$\therefore$ Weight of NH_3 per 100 g water in saturated solution at 25°C and P mm Hg

$$= \frac{W_2 \times 100}{(W_1 - W) - W_2} \text{ g} = \underline{\hspace{2cm}} \text{ g}$$

	Water	1/20 NaOH	1/20 NaOH	1/10 NaOH
	$Ca(OH)_2$	$Ca(OH)_2$	$Ca(OH)_2$	$Ca(OH)_2$
Burette (final)				
Burette (initial)				
Titre (ml)				
Mean titre (ml)				

EXPERIMENT 3. TO DETERMINE THE SOLUBILITY PRODUCT OF CALCIUM HYDROXIDE BY A TITRATION METHOD

APPARATUS AND CHEMICALS REQUIRED

Four 150- or 250-ml stoppered bottles; four dry filter funnels; five dry 250-ml conical flasks; burette and stand; white tile; 10-ml pipette; 25-ml pipette; 50-ml pipette; filter papers; thermostat (if available); 0–110°c (in 1°c) thermometer; small labels; three 100-ml graduated flasks; wash bottle.

Solid calcium hydroxide (8 g); 0·1M (0·1N) sodium hydroxide (100 ml); 0·1M (0·1N) hydrochloric acid (250 ml); phenolphthalein indicator.

EXPERIMENTAL PROCEDURE

Prepare 100 ml each of N/20, N/40 and N/100 sodium hydroxide by diluting appropriate pipetted volumes of the N/10 sodium hydroxide.

Label four dry, stoppered, bottles and place about 2 g of calcium hydroxide in each. Into the first place about 100 ml distilled water, into the second about 100 ml N/20 alkali, into the third N/40 alkali and into the fourth N/100 alkali, again using about 100 ml in each case. Shake well and place in a 25°c thermostat for at least 30 min (shake well from time to time during this period). If a thermostat is not available, then allow to attain equilibrium at room temperature, which must be noted.

Filter the contents of each bottle (using a fresh filter paper for each) into a separate dry conical flask, rejecting the first 5 ml (approx.) of filtrate in each case. (The first few ml are rejected because they are less concentrated in solute than the remainder, the filter paper adsorbing solute until it attains an equilibrium with the solution.) Titrate two separate 25-ml portions of each filtrate with 0·1N hydrochloric acid, using phenolphthalein indicator.

Calculate the solubility product of calcium hydroxide from the results of each equilibration. Estimate the probable error in the final results, and then express them to an appropriate order of accuracy.

RESULTS

Titration of solutions saturated with calcium hydroxide

In burette: 0·1N hydrochloric acid.   Indicator: phenolphthalein.
25 ml of filtrate used per titration.

	Water + Ca(OH)$_2$		N/20 NaOH + Ca(OH)$_2$		N/40 NaOH + Ca(OH)$_2$		N/100 NaOH + Ca(OH)$_2$	
Burette (finish)								
Burette (start)								
Titre (ml)								
Mean titre (ml)								

From these titration results calculate the normalities of the filtrates with respect to total $OH^-(Ca(OH)_2 + NaOH)$ in each case.

Complete the following results table:

Ca(OH)$_2$ solution prepared in	Total OH$^-$ ion conc. (g.ions/litre)	Conc. OH$^-$ from lime only (g.ions/litre)	Conc. Ca^{2+} (g.ions/litre)	Solubility product [Ca^{2+}][OH$^-$]^2
Water	N_1			
N/20 NaOH	N_2			
N/40 NaOH	N_3 $\Big\}$ N_x			
N/100 NaOH	N_4			

Thermostat (or room, as applicable) temperature = _____ °C

THEORY

For an aqueous solution of calcium hydroxide in contact with the solid there exists the equilibrium

$$\underset{\text{solid}}{Ca(OH)_2} \rightleftharpoons \underset{\text{in solution}}{Ca^{2+} + 2OH^-}$$

then, at constant temperature, by the law of chemical equilibrium

$$\frac{[Ca^{2+}][OH^-]^2}{[Ca(OH)_2]} = constant_1$$

and since, for heterogeneous equilibria, the concentration of solid phases may be taken as constant

$$[Ca^{2+}][OH^-]^2 = constant_2$$

and this constant$_2$ is the 'solubility product' of calcium hydroxide at this temperature.

In this experiment the titrations determine the total OH$^-$ ion concentration. With no sodium hydroxide present and all the hydroxyl ion being supplied by the slaked lime,

$$\text{g.ionic concentration } OH^- = N_1$$
$$\therefore \text{ g.ionic concentration } Ca^{2+} = N_1/2$$

In the presence of sodium hydroxide of normality N_y the total normality of the alkaline solution N_x (determined by titration) is due to OH$^-$ ions from both the slaked lime and the sodium hydroxide, hence

$$\text{g.ionic concentration } OH^- \text{ from } Ca(OH)_2 = N_x - N_y$$
$$\therefore \text{ g.ionic concentration } Ca^{2+} = \tfrac{1}{2}(N_x - N_y)$$

The Partition Law

EXPERIMENT 4. TO DETERMINE THE PARTITION COEFFICIENT FOR AMMONIA BETWEEN WATER AND CHLOROFORM

APPARATUS AND CHEMICALS REQUIRED

Two burettes and stands; white tile; two 25-ml pipettes; stoppered bottle; 250-ml conical flask; retort stand and clamp; pipette safety filler; two 250-ml beakers; wash bottle; 0–110°c (in 1°c) thermometer; thermostat (if available).

Approx. 1·0M (1·0N) ammonia (150 ml); 0·5M (1·0N) sulphuric acid (150 ml); 0·025M (0·05N) sulphuric acid (150 ml); methyl red indicator; chloroform (60 ml).

EXPERIMENTAL PROCEDURE

Pipette 50 ml of approx. 1·0N ammonia and 50 ml chloroform (dry pipette) into a stoppered bottle, shake well for 2 or 3 min and then put aside (preferably in a thermostat at 25°c) for about 15 min to attain equilibrium. Note the thermostat (or room, as applicable) temperature.

While the mixture is attaining equilibrium, standardise the approx. 1·0N ammonia by titrating with 1·0N sulphuric acid using methyl red indicator.

Analyse the chloroform layer for ammonia by titrating a 25-ml portion of it with 0·05N sulphuric acid, using methyl red indicator. To achieve an accurate result the following sampling and titrating procedure should be adopted:

Close the dry pipette with the finger and carefully lower the tip into the bottle until it is about three-quarters of the way into the chloroform bulk and well away from the interface and from glass surfaces. (This operation is facilitated by clamping the pipette to a retort stand.) Gently blow through the pipette, to expel any small quantity of the aqueous layer which may have entered, and then withdraw the sample and run into a conical flask. Add about 25 ml distilled water to 'extract' the ammonia from the chloroform and then titrate. Shake the flask frequently during the titration. There is a considerable tendency for the red (red in aqueous solution, that is) form of the indicator to dissolve preferentially in the chloroform layer to give a yellow solution, and a little more indicator may have to be added as the equivalence-point is approached. Because of the yellow background produced in this way the equivalence-point must be observed horizontally through the aqueous layer.

Tabulate results and calculate the normality of the ammonia in the chloroform layer. From this and the determined normality of the original aqueous ammonia calculate the partition coefficient for ammonia between water and chloroform. Estimate the probable error in the final result and then express it to an appropriate order of accuracy.

RESULTS

(1) Titration of approx. 1·0N ammonia with 1·0N sulphuric acid

25 ml ammonia solution used per titration.   Indicator: methyl red.

8

Burette (finish)		
Burette (start)	--------	--------
Titre (ml)	7 · 8	

_____ ml 1·0N sulphuric acid ≡ 25 ml ammonia solution

(2) Titration of chloroform layer with 0·05N sulphuric acid

25 ml chloroform layer plus approx. 25 ml distilled water   Indicator: methyl red.
used per titration.

Burette (finish)	
Burette (start)	--------
Titre (ml)	

_____ ml 0·05N sulphuric acid ≡ 25 ml ammoniated chloroform

Thermostat (or room, as applicable) temperature = _____°C

CALCULATION

Normality of original ammonia solution   = _____ = _____N (N_0)

Normality of ammonia in chloroform layer at equilibrium = _____ = _____N (N_c)

As equal volumes of N_0 ammonia and chloroform were equilibrated:

Normality of aqueous layer at equilibrium = $N_0 - N_c = N_w$

Partition coefficient = $\dfrac{\text{Equilibrium conc. ammonia in aqueous layer}}{\text{Equilibrium conc. ammonia in chloroform layer}}$

$$= \frac{N_w}{N_c} = \frac{N_0 - N_c}{N_c} \quad \text{at} \underline{\quad}°C$$

THEORY

Nernst's Partition Law (Distribution Law) states:
'The ratio of the concentrations of a solute distributed between two given immiscible solvents which are in contact and at equilibrium is constant, at constant temperature, providing the molecular state is the same in both solvents.'

The constant ratio of the concentrations is known as the partition (or distribution) coefficient for the system.

Distribution phenomena have important applications in solvent/solvent extraction methods in organic (and to some extent inorganic) separations and in the investigation of some chemical equilibria (Experiments 6 and 7).

EXPERIMENT 5. TO DETERMINE THE PARTITION COEFFICIENT FOR ACETIC ACID BETWEEN WATER AND CARBON TETRACHLORIDE

APPARATUS AND CHEMICALS REQUIRED

Three stoppered bottles; two 100-ml measuring cylinders; two 25-ml pipettes; 5-ml graduated pipette; two burettes and stands; white tile; pipette safety filler; three conical flasks (250 ml); wash bottle; thermostat (if available); 0–110°C (in 1°C) thermometer; retort stand and clamp.

Glacial acetic acid (10 ml); carbon tetrachloride (225 ml); 1·0M (1·0N) sodium hydroxide (250 ml); M/40 (N/40) sodium hydroxide (250 ml); phenolphthalein indicator.

EXPERIMENTAL PROCEDURE

Into each of three dry, stoppered bottles place 75 ml water and 75 ml carbon tetra-chloride (use measuring cylinders). Label the bottles A, B and C. Add 2 ml of glacial acetic acid to A, 3 ml to B, and 4 ml to C. Shake well and put aside for about 15 min (preferably in a thermostat at 25°C) to attain equilibrium. Shake the bottles two or three times during this equilibration period. Charge one burette with 1·0N sodium hydroxide and the other with N/40 sodium hydroxide. Label the burettes.

From A withdraw two separate 25-ml portions of the aqueous layer and run into conical flasks. (The withdrawal is facilitated by clamping the pipette on a retort stand, so that it may be lowered accurately into the aqueous layer.) Titrate the aqueous portions with 1·0N sodium hydroxide, using phenolphthalein indicator. Now withdraw two 25-ml portions of the carbon tetrachloride layer into conical flasks, add about 10 ml distilled water to each (to 'extract' the acid from the carbon tetrachloride) and then titrate with N/40 sodium hydroxide, using phenolphthalein indicator.

N.B.

When withdrawing and titrating the carbon tetrachloride layer the following precautions must be taken:

(a) Ensure that the tip of the pipette remains near the centre of the carbon tetrachloride layer, *i.e.* well away from the interface and from the glass surfaces.

(b) Before drawing the sample into the pipette blow gently to expel any small quantity of the aqueous layer that may have entered during the immersion of the tip.

(c) During the titration the flask should be tightly corked and shaken vigorously from time to time, particularly as the equivalence-point is approached.

Analyse the contents of bottles B and C in a similar way.

Record the thermostat (or room) temperature.

Tabulate results and calculate the partition coefficient for acetic acid between water and carbon tetrachloride from each set of results. Estimate the probable error in the final results and then express them to an appropriate order of accuracy.

10

RESULTS

Quantity of acetic acid in mixture	Titre of N NaOH for 25 ml of aqueous layer (x ml)	Titre of N/40 NaOH for 25 ml of CCl₄ layer (y ml)	$\dfrac{x}{y/40}$	$\dfrac{x}{\sqrt{y/40}}$
2 ml	(i)	(i)		
	(ii)	(ii)		
	Mean	Mean		
3 ml	(i)	(i)		
	(ii)	(ii)		
	Mean	Mean		
4 ml	(i)	(i)		
	(ii)	(ii)		
	Mean	Mean		

Thermostat (or room, as applicable) temperature = _____ °C

THEORY

First consider Nernst's Partition Law as stated in Experiment 4.

In the case of acetic acid distributed between water and carbon tetrachloride, the ratio of the concentrations is not constant from one mixture to another. A close approximation to constancy is, however, given by the ratio

$$\frac{\text{Conc. of acetic acid in water}}{\sqrt{\text{Conc. of acetic acid in carbon tetrachloride}}}$$

and this may be explained if one assumes almost complete dimerisation in the organic layer.

Consider the system in which a solute X is distributed between two immiscible solvents A and B. In A the solute is monomeric but in B it is mostly associated as X_n. There will exist two equilibria, one across the interface and one within solvent B:

$$\begin{array}{ll} X & \text{solvent A} \\ \text{—⇅—————} & \\ nX \rightleftharpoons X_n & \text{solvent B} \end{array}$$

For the equilibrium within B

$$\frac{[X_n]_B}{[X]_B{}^n} = a \text{ const.} \quad \text{or} \quad [X]_B{}^n = k_1[X_n]_B \quad . \quad . \quad . \quad . \quad . \quad \text{(i)}$$

For the equilibrium across the interface

$$\frac{[X]_B}{[X]_A} = a \text{ const.} \quad \text{or} \quad [X]_B = k_2[X]_A \qquad \qquad \text{(ii)}$$

From (i) and (ii)

$$[X]_B{}^n = k_1[X_n]_B = k_2{}^n[X]_A{}^n$$

$$\text{i.e.} \quad \frac{[X]_A{}^n}{[X_n]_B} = k_3 \quad \text{or} \quad \frac{[X]_A}{\sqrt[n]{[X_n]_B}} = k_4$$

Providing the association in solvent B is almost complete, $[X_n]_B$ may be taken as the total concentration of solute in B, and it is this which is measured during the analysis of the solution.

Hence

$$\frac{[X]_A}{\sqrt[n]{[X]_B}} = k_5$$

In the case of acetic acid between water and carbon tetrachloride, $n=2$, which indicates dimerisation of the acetic acid in the organic solvent.

There is a slight deviation from constancy in the results, thus calculated, in this experiment because this theory takes no account of

- (a) the small quantity of monomeric acid in the carbon tetrachloride,
- (b) the slight (and varying with conc.) dissociation of the acid in the aqueous layers,
- (c) the fact that the solvents are not completely immiscible and the degree of miscibility of the solvents varies with the concentration of the common solute, *e.g.* water is more soluble in carbon tetrachloride which contains acetic acid than in pure carbon tetrachloride.

EXPERIMENT 6. TO DETERMINE THE FORMULA OF THE CUPRAMMONIUM ION BY A PARTITION METHOD

APPARATUS AND CHEMICALS REQUIRED

Two burettes and stands; white tile; two 25-ml pipettes; 50-ml pipette; two 250-ml stoppered bottles; 250-ml conical flask; retort stand and clamp; pipette safety filler; three 250-ml beakers; wash bottle; 0–110°C (in 1°C) thermometer; thermostat if available.

Approx. 1·0M (1·0N) ammonia (150 ml); 0·5M (1·0N) sulphuric acid (100 ml); 0·025M (0·05N) sulphuric acid (100 ml); 0·1M (0·1N) copper sulphate (6·238 g of uneffloresced pentahydrate in 250 ml); chloroform (75 ml); methyl red indicator.

EXPERIMENTAL PROCEDURE

Determine the partition coefficient for ammonia between water and chloroform as in Experiment 4. If the partition coefficient has not already been determined the two experiments may be run concurrently.

Pipette 25 ml of approximately 1·0N ammonia, 25 ml of 0·1M copper sulphate and 75 ml chloroform (dry pipette) into a stoppered bottle. Shake well for 2 or 3 min and then put aside (preferably in a thermostat) for about 15 min to attain equilibrium. Shake the bottle two or three times during this equilibration period.

While the mixture is attaining equilibrium, standardise the approx. 1·0N ammonia by titrating with 1·0N sulphuric acid using methyl red indicator.

Carefully withdraw 50 ml of the chloroform layer into a conical flask, add about 25 ml distilled water and then titrate with 0·05N sulphuric acid, using methyl red indicator.

N.B.

Adopt the same sampling and titrating procedure as in the partition coefficient determination in Experiment 4.

From the results of the above titration and the determined partition coefficient calculate the value of x in the formula $Cu(NH_3)_x^{2+}$.

RESULTS AND CALCULATION

Normality of original ammonia solution = _____N (N_0)

Partition coeff. NH_3 between water and chloroform = _____ (P)

Titration of chloroform layer of the cuprammonium–chloroform mixture

50 ml of chloroform layer and approx. 25 ml distilled water in flask.   Indicator: methyl red.

Burette (finish)	
Burette (start)	
Titre (ml)	

_____ ml 0·05N sulphuric acid $\equiv$ 50 ml ammoniated chloroform

Normality of chloroform layer $\qquad\qquad\qquad\qquad$ = _____ N (N_1)

$\therefore$ Normality free ammonia in cuprammonium layer $\qquad$ = $P.N_1$

Moles free ammonia originally pipetted into bottle $\qquad$ = $\dfrac{25}{1000} N_0$

Moles free ammonia in cuprammonium layer at equilibrium = $\dfrac{50}{1000} P.N_1$

Moles free ammonia in chloroform layer at equilibrium $\quad$ = $\dfrac{75}{1000} N_1$

g.ions Cu^{2+} pipetted into bottle $\qquad\qquad\qquad$ = $\dfrac{25 \times 0.1}{1000}$

$\therefore$ Moles NH_3 taken up to form $Cu(NH_3)_x^{2+}$ $\qquad$ = $\dfrac{2.5x}{1000}$

$$\therefore \frac{25}{1000} N_0 = \frac{50}{1000} P.N_1 + \frac{75}{1000} N_1 + \frac{2.5x}{1000}$$

$$\therefore N_0 = 2P.N_1 + 3N_1 + 0.1x$$

$$\therefore x = 10(N_0 - 2P.N_1 - 3N_1)$$

$$= \underline{\hspace{3cm}}$$

THEORY

The equilibrium

$$Cu^{2+} + xNH_3 \rightleftharpoons Cu(NH_3)_x^{2+}$$

lies well to the right and in the presence of excess ammonia the complexing of the cupric ions is almost complete.

If the cuprammonium solution is shaken with chloroform the free ammonia becomes distributed between the two layers according to the Partition Law. A sample of the chloroform layer may be withdrawn and analysed for ammonia without disturbing the main equilibrium. If the concentration of the ammonia in the chloroform layer is determined and the partition coefficient for ammonia between water and chloroform is known, then the concentration of free ammonia in the cuprammonium layer may be calculated. If the original quantity of ammonia placed in the mixture is known, then that which has been complexed with the cupric ions may be found by difference. Knowing the original cupric ion concentration, x may therefore be calculated.

EXPERIMENT 7. TO DETERMINE THE EQUILIBRIUM CONSTANT FOR THE REACTION $I_3^- \rightleftharpoons I_2 + I^-$ BY A PARTITION METHOD

APPARATUS AND CHEMICALS REQUIRED

Three 250-ml stoppered bottles; 5-ml and 25-ml pipettes; 10-ml microburette; white tile; four conical flasks (250 ml); filter funnel; safety pipette filler; thermostat; two 250-ml measuring cylinders; 0–110°C (in 1°C) thermometer, burette stand.

Iodine (1 g); carbon tetrachloride (60 ml); 0·1M (0·1N) potassium iodide (200 ml); 0·1M (0·1N) sodium thiosulphate (200 ml); 10% potassium iodide (50 ml); starch indicator.

EXPERIMENTAL PROCEDURE

Dissolve about 1 g of crushed iodine in 60 ml of carbon tetrachloride in a clean, dry conical flask, and then filter into a dry, stoppered bottle.

Pipette 25 ml of the iodine–carbon tetrachloride solution into each of two clean, stoppered bottles and to one add about 200 ml of water and to the other about 200 ml of 0·1N potassium iodide. Label the bottles appropriately, shake well for about 1 min, and place aside (preferably in a thermostat at 25°C) for 15–30 min to attain equilibrium. Shake the bottles two or three times during this equilibration period. Record the room or thermostat temperature as applicable.

Analyse the contents of each bottle in the following way. Carefully withdraw 50 ml of the aqueous layer and 5 ml of the organic layer into separate conical flasks and titrate each with 0·1N thiosulphate, using starch indicator. Before titrating the organic sample add about 10 ml of 10% potassium iodide solution to ensure complete 'extraction' of the iodine from the carbon tetrachloride. Titrate slowly with care.

Tabulate results and calculate the equilibrium constant K. Estimate the probable error in the final result and then express the result to an appropriate order of accuracy.

RESULTS

0·1N sodium thiosulphate in the burette and starch indicator used for all titrations.

Iodine–carbon tetrachloride–water equilibration

50 ml aqueous layer used per titration

Burette (finish)	
Burette (start)	
Titre (ml)	V_1

50 ml aqueous layer $\equiv V_1$ ml 0·1N thiosulphate

5 ml carbon tetrachloride layer used per titration

Burette (finish)	---------
Burette (start)	
Titre (ml)	V_2

5 ml of carbon tetrachloride layer $\equiv V_2$ ml 0·1N thiosulphate

Iodine–carbon tetrachloride–potassium iodide equilibration

50 ml aqueous potassium iodide layer used per titration

Burette (finish)	---------
Burette (start)	
Titre (ml)	V_3

50 ml aqueous KI layer $\equiv V_3$ ml 0·1N thiosulphate

5 ml carbon tetrachloride layer used per titration

Burette (finish)	---------
Burette (start)	
Titre (ml)	V_4

5 ml carbon tetrachloride layer $\equiv V_4$ ml 0·1N thiosulphate

Thermostat (or room, as applicable) temperature = _____°C

CALCULATION

(1) To calculate the partition coefficient for I_2 between water and carbon tetrachloride from the results of first equilibration.

Normality iodine in aqueous layer $\qquad = \dfrac{V_1}{50} 0·1$

Normality iodine in CCl_4 layer $\qquad = \dfrac{V_2}{5} 0·1$

$\therefore$ Partition coeff. I_2 between water and $CCl_4 = \dfrac{5V_1}{50V_2} = \dfrac{V_1}{10V_2}$

(2) To calculate the equilibrium constant K from the results of the second equilibration and the above partition coefficient.

Normality of iodine in aq. KI $= \dfrac{V_3}{50} 0 \cdot 1$ (*i.e.* free I_2 and the I_2 in $I_3{}^-$)

Normality I_2 in CCl_4 layer $= \dfrac{V_4}{5} 0 \cdot 1$

Normality **free iodine** in KI layer $= \dfrac{V_4}{50} \times$ (partition coefficient)

$$= \frac{V_4 V_1}{500 V_2} = N_1$$

Normality $I_3{}^-$ in KI layer $= \dfrac{V_3}{50} 0 \cdot 1 - \dfrac{V_4 V_1}{500 V_2}$

$$= N_2$$

Hence, normality I^- in KI layer $= 0 \cdot 1 - 0 \cdot 5 * \left[\dfrac{V_3}{50} 0 \cdot 1 - \dfrac{V_4 V_1}{500 V_2} \right]$

$$= N_3$$

Converting to molarities

$$K = \frac{[I_2][I^-]}{[I_3{}^-]} = \frac{0 \cdot 5 N_1 \times N_3}{0 \cdot 5 N_2} = \frac{N_1 N_3}{N_2}$$

THEORY

When iodine dissolves in potassium iodide solution the ion $I_3{}^-$ is formed and is in equilibrium with iodide ions and free iodine according to the equation

$$I_3{}^- \rightleftharpoons I_2 + I^-$$

If this equilibrium is set up in contact with, say, carbon tetrachloride, then the iodine in solution in the organic layer is in equilibrium with the *free* iodine in the potassium iodide solution according to the Partition Law. Thus, providing the partition coefficient for iodine between water and carbon tetrachloride is known, analysis of the organic layer for iodine permits the calculation of the free iodine concentration in the potassium iodide solution. If the original concentration of potassium iodide is known and the total tri-iodide and free iodine in the aqueous layer determined by titration, the equilibrium constant may be calculated.

*Factor of 0.5 Since E.W. $KI_3 = \frac{1}{2}$M.W. but E.W. KI = M.W.

Thermochemistry

The calculations in this section are performed in a strictly algebraic manner. The student must allow signs to take care of themselves. A negative value for any heat quantity indicates that it is heat given up or evolved ((−) = exothermic), and a positive value indicates that it is heat absorbed ((+) = endothermic).

EXPERIMENT 8. THE DETERMINATION OF HEATS OF NEUTRALISATION

APPARATUS AND CHEMICALS REQUIRED

Small vacuum flask (not less than 150-ml capacity) fitted with a *cork* and 0–50°C (in 0·1°C) thermometer; 400-ml beaker; 50-ml pipette; several clean boiling tubes and rack; tripod; bunsen.

1·0M (1·0N) sodium hydroxide (125 ml); 1·0M (1·0N) hydrochloric acid (75 ml); 1·0M (1·0N) nitric acid (75 ml).

EXPERIMENTAL PROCEDURE

(1) Determination of the water equivalent of the vacuum flask

Two-thirds fill the beaker with tap water and bring nearly to the boil. Dry the inside of the flask and pipette into it 50 ml of distilled water, place the cork and thermometer in position, shake, and note the steady temperature. Pipette 50 ml distilled water into a clean, dry, boiling tube and immerse the tube in the hot water until the distilled water is at 45–50°C. Remove the tube from the water and note the temperature of the distilled water after stirring well with the thermometer. Immediately pour the 50 ml of distilled water into the flask, close with cork and thermometer, shake, and note the steady temperature.

Calculate the water equivalent of the vacuum flask.

(2) Determination of heat of neutralisation

Drain the vacuum flask and pipette into it 50 ml of 1·0N sodium hydroxide. Place the cork and thermometer in position and note the steady temperature. Pipette 50 ml of 1·0N hydrochloric acid into a boiling tube and record its temperature. Pour the 50 ml of acid into the flask with the alkali, close with cork and thermometer, shake, and note the steady temperature.

Repeat using 50 ml of 1·0N nitric acid.

Calculate the heats of neutralisation of sodium hydroxide by hydrochloric and nitric acids. Estimate the probable error in the final results and then express them to an appropriate order of accuracy.

RESULTS AND CALCULATION

First read the note at the top of page 18.

(1) Determination of water equivalent

Initial temperature of flask + 50 ml cold water $\quad = t_1°C$

Initial temperature of hot water $\quad\quad\quad\quad\quad = t_2°C$

Final temperature of flask and water $\quad\quad\quad = t_3°C$

Weight of hot water added to flask (50 ml at room temp.) = 50 g

Let W g be water equivalent of the vacuum flask. Then, since there is no chemical reaction,

$$\frac{\text{Change in heat}}{\text{content of flask}} + \frac{\text{Change in heat content}}{\text{of cold water}} + \frac{\text{Change in heat content}}{\text{of hot water}} = 0$$

i.e.
$$W(t_1 - t_3) + 50(t_1 - t_3) + 50(t_2 - t_3) = 0$$

$$W = \frac{50(t_2 - t_3)}{(t_3 - t_1)} - 50 \text{ g}$$

Label the flask with this value for future reference.

(2) Determination of heat of neutralisation

Initial temperature of flask and alkali $\quad = T_1°C$

Initial temperature of acid $\quad\quad\quad\quad = T_2°C$

Volume of $1\cdot0$N sodium hydroxide in flask $= 50$ ml

Volume of $1\cdot0$N hydrochloric acid added $\quad = 50$ ml

Final temperature of flask and solution $\quad = T_3°C$

Water equivalent of vacuum flask $\quad\quad = W$ g

In this calculation it is assumed that the thermal capacity of 50 ml of any one of the solutions is approximately equal to the thermal capacity of 50 g of water. This is not an unreasonable assumption since the slightly decreased specific heat is almost offset by the slightly higher mass.

Let the heat change during this neutralisation be Q calories. Hence

$$Q = W(T_1 - T_3) + 50(T_1 - T_3) + 50(T_2 - T_3) \text{ cals}$$

Now 50 ml of a normal solution contain 1/20th gram equivalent. Therefore, heat of neutralisation of sodium hydroxide by hydrochloric acid is equal to $20Q$ calories.

Similarly for the reaction using nitric acid.

THEORY

The heat of neutralisation of an acid by a base is the heat evolved when one gram equivalent of the acid is neutralised by one gram equivalent of the base.

The values obtained in this experiment for the reactions

$$HCl + NaOH \rightarrow NaCl + H_2O$$
$$HNO_3 + NaOH \rightarrow NaNO_3 + H_2O$$

will be approximately the same, since in both cases the essential reaction is

$$H^+ + OH^- \rightarrow H_2O$$

and the reactants, being strong electrolytes, may be assumed completely dissociated.

For temperature measurements in experiments of this type a pair of thermometers which have been calibrated against each other should preferably be used. Failing this the same thermometer should be used for all temperature measurements during any one experiment.

EXPERIMENT 9. THE DETERMINATION OF THE HEAT OF HYDRATION OF ANHYDROUS SODIUM CARBONATE

APPARATUS AND CHEMICALS REQUIRED

Small vacuum flask (not less than 150-ml capacity) fitted with a *cork* and 0–50°c (in 0·1°c) thermometer; 250-ml beaker; 50-ml pipette; several boiling tubes and rack; desiccator; evaporating basin; tripod; bunsen; tongs; 10-ml graduated pipette.

Sodium carbonate decahydrate (12 g) (a sample which has not effloresced is essential); sodium bicarbonate (10 g).

EXPERIMENTAL PROCEDURE

Ignite about 10 g of sodium bicarbonate in an evaporating basin to convert it to anhydrous sodium carbonate (30 min). Place this in a desiccator to cool.

If the water equivalent of the vacuum flask is not already known, determine this as in the first part of Experiment 8.

Weigh out about 12 g (to 0·01 g) of the hydrated salt, place this in the dry vacuum flask and replace the cork and thermometer. Note the steady temperature when equilibrium is attained. Two-thirds fill the beaker with tap water and bring nearly to the boil. Pipette 50 ml of distilled water into a clean, dry boiling tube and immerse this in the hot water until the temperature of the distilled water is 45–50°c. Remove the tube from the hot water bath and note the temperature of the distilled water after stirring well with the thermometer. Immediately pour the distilled water on to the hydrate in the flask, close with bung and thermometer, shake well and note the steady temperature. Wash and **thoroughly dry** the flask.

Weigh out between 4–5 g (to 0·01 g) of the anhydrous salt and repeat the above procedure. The water this time should be of volume 57·5 ml and the temperature not below 40°c. (The larger volume of water ensures that the composition of the solution produced is identical with that of the hydrate in the first determination, and this is to eliminate a large error which would otherwise be introduced owing to the considerable endothermic heat of dilution. The temperature of the added water should be several degrees above the transition temperature for the hydrate–anhydrous salt system; otherwise the salt rapidly sets to a hard mass and then dissolves very slowly.)

Calculate the heats of solution of the decahydrate and the anhydrous salt, and from these calculate the heat of hydration of the anhydrous salt. Estimate the probable error in the final result and then express this result to an appropriate order of accuracy.

RESULTS AND CALCULATION

First read the note at the top of page 18.

(1) For the heat of solution of the decahydrate

Weight of hydrate $= w_1$ g

Weight of added water (50 ml at room temperature) $= 50$ g

Initial temperature of flask and hydrate $= t_1 °C$

Initial temperature of added water $= t_2 °C$

Final temperature of flask and solution $= t_3 °C$

Water equivalent of flask $= W$ g

Molecular weight of $Na_2CO_3,10H_2O$ $= 286$

In this calculation the thermal capacity of the anhydrous salt is neglected, but the considerable thermal capacity of the water of crystallisation of the hydrate is taken into account.

Heat of soln.	Change in	Change in	Change in heat
of small quantity =	heat content +	heat content of +	content of
of hydrate	of flask	water of cryst.	added water

i.e.
$$Q_1 = W(t_1 - t_3) + \frac{180}{286} w_1(t_1 - t_3) + 50(t_2 - t_3) \text{ cals}$$

Hence the molar heat of solution

$$\Delta H_1 = \frac{286}{w_1} Q_1 \text{ cals}$$

or
$$\Delta H_1 = \frac{286}{w_1} \left[W(t_1 - t_3) + \frac{180}{286} w_1(t_1 - t_3) + 50(t_2 - t_3) \right] \text{ cals}$$

Heat of solution of sodium carbonate decahydrate = _____ cals

(2) For the heat of solution of the anhydrous salt

Weight of anhydrous salt $= w_2$ g

Weight of added water (57·5 ml at room temperature) $= 57·5$ g

Initial temperature of flask and anhydrous salt $= T_1 °C$

Initial temperature of the added water $= T_2 °C$

Final temperature of flask and solution $= T_3 °C$

Water equivalent of flask $= W$ g

Molecular weight of Na_2CO_3 $= 106$

Neglecting the thermal capacity of the anhydrous salt

Heat of solution	Change in heat	Change in heat
of small quantity =	content +	content of
of anhydrous salt	of flask	added water

i.e.
$$Q_2 = W(T_1 - T_3) + 57·5(T_2 - T_3) \text{ cals}$$

Hence molar heat of solution

$$\Delta H_2 = \frac{106}{w_2} Q_2 \text{ cals}$$

or

$$\Delta H_2 = \frac{106}{w_2} \left[W(T_1 - T_3) + 57{\cdot}5(T_2 - T_3) \right] \text{ cals}$$

Heat of solution of anhydrous sodium carbonate = _____ cals

(3) For the heat of hydration of the anhydrous salt

$$\text{Na}_2\text{CO}_3,10\text{H}_2\text{O(s)} + \text{Aq}_1 \rightarrow \text{Na}_2\text{CO}_3(\text{Aq}_2) + \Delta H_1 \text{ cals} \quad . \quad . \quad . \quad . \quad \text{(i)}$$

$$\text{Na}_2\text{CO}_3(s) + \text{Aq}_2 \rightarrow \text{Na}_2\text{CO}_3(\text{Aq}_2) + \Delta H_2 \text{ cals} \quad . \quad . \quad . \quad . \quad \text{(ii)}$$

The volumes of water used are adjusted such that

$$\text{Aq}_2 - \text{Aq}_1 = 10\text{H}_2\text{O (l)}$$

Therefore, by Hess's law, subtracting (ii) − (i) gives

$$\text{Na}_2\text{CO}_3(s) + 10\text{H}_2\text{O(l)} \rightarrow \text{Na}_2\text{CO}_3 . 10\text{H}_2\text{O(s)} + \Delta H_2 - \Delta H_1$$

Heat of hydration of anhydrous sodium carbonate is

$$\Delta H_2 - \Delta H_1 = \text{_____} \text{ cals}$$

EXPERIMENT 10. TO DETERMINE THE HEAT OF THE SECOND IONISATION OF CARBONIC ACID

APPARATUS AND CHEMICALS REQUIRED

Small vacuum flask (not less than 150-ml capacity) fitted with a *cork* and 0–50°c (in 0·1°c) thermometer; 400-ml beaker; 50-ml pipette; several clean, dry, boiling tubes and rack; tripod; bunsen.

1·0M (1·0N) sodium hydroxide (75 ml); 1·0M (1·0N) hydrochloric acid (75 ml); 1·0M (1·0N) sodium bicarbonate (75 ml).

EXPERIMENTAL PROCEDURE

First determine the water equivalent of the vacuum flask and the heat of neutralisation of sodium hydroxide by hydrochloric acid (ΔH_1), both as in Experiment 8.

Repeat the heat of neutralisation experiment using 50 ml of 1·0N sodium bicarbonate in place of the *hydrochloric acid* (ΔH_2).

Calculate ΔH for the second ionisation of carbonic acid from the results obtained. Estimate the probable error in the final result and then express this result to an appropriate order of accuracy.

RESULTS

(1) Determination of the water equivalent of the vacuum flask

See Experiment 8, p. 18.

(2) Determination of the heat of neutralisation of sodium hydroxide by hydrochloric acid (ΔH_1)

See Experiment 8, p. 18.

(3) Determination of the heat of the reaction between sodium hydroxide and sodium bicarbonate (ΔH_2)

Set out results as for NaOH/HCl reaction.

The reactions of which the heats have been determined are:

NaOH/HCl $\qquad$ $H^+ + OH^- \to H_2O$ $\qquad$ (ΔH_1) . . . . (i)

NaOH/NaHCO$_3$ $\quad$ $HCO_3^- + OH^- \to H_2O + CO_3^{2-}$ (ΔH_2) . . . . . . (ii)

By Hess's law, subtracting (ii) − (i),

$$HCO_3^- \to H^+ + CO_3^{2-} \quad (\Delta H = \Delta H_2 - \Delta H_1)$$

i.e. the heat of the second ionisation of carbonic acid

$$\Delta H = \Delta H_2 - \Delta H_1 = \underline{\qquad} \text{ cals}$$

Buffer Solutions

EXPERIMENT 11. PREPARATION OF SOME BUFFER SOLUTIONS AND THEIR USE TO DETERMINE THE WORKING RANGES OF SOME pH INDICATORS

The buffer solutions prepared in this experiment are in the range pH = 1 to pH = 7 and are required also for Experiments 12 and 13, and so should be carefully preserved.

APPARATUS AND CHEMICALS REQUIRED

50 clean test-tubes of about the same bore, together with sufficient racks to accommodate them; 25 small (60-ml) stoppered bottles (or corked boiling tubes in racks will suffice);

Buffer ref. no.	pH	Citrate soln. (ml)	0·1N HCl (ml)
1	1·17	5·0	45·0
2	1·43	10·0	40·0
3	1·93	15·0	35·0
4	2·27	16·65	33·35
5	2·97	20·0	30·0
6	3·36	22·5	27·5
7	3·53	23·75	26·25
8	3·69	25·0	25·0
9	3·95	27·5	22·5
10	4·16	30·0	20·0
11	4·45	35·0	15·0
12	4·65	40·0	10·0
13	4·83	45·0	5·0
14	4·89	47·5	2·5
15	4·96	50·0	0

Buffer ref. no.	pH	Citrate soln. (ml)	0·1N NaOH (ml)
16	5·02	47·5	2·5
17	5·11	45·0	5·0
18	5·31	40·0	10·0
19	5·57	35·0	15·0
20	5·98	30·0	20·0
21	6·34	27·5	22·5
22	6·69	26·25	23·75

three burettes and stands; 1000-ml graduated flask; 100-ml pipette; 100-ml beaker; dropping pipette with teat; wash bottle; three filter funnels; three 250-ml beakers.

Citric acid (pure cryst.) (21 g); 1·0M (1·0N) sodium hydroxide (250 ml); 0·1M (0·1N) sodium hydroxide (150 ml); 0·1M (0·1N) hydrochloric acid (400 ml); methyl orange indicator (screened or unscreened); methyl red indicator.

EXPERIMENTAL PROCEDURE

Weigh out accurately (in a 100-ml beaker) 21·00 g (to 0·01 g) of citric acid and wash this into a 1000-ml graduated flask with distilled water. Pipette into the flask 200-ml of 1·0N sodium hydroxide and swirl to dissolve the acid. Make up to the mark with distilled water and shake well to ensure homogeneity. (This prepares a solution of disodium hydrogen citrate.)

Prepare the buffer solutions (page 25) in clean, well-drained reagent bottles by mixing the citrate solution with either 0·1N hydrochloric acid or 0·1N sodium hydroxide as specified. Use burettes to measure the required volumes. Carefully label each buffer solution with a reference number and its pH.

In each of 22 clean, dry test-tubes place approx. 5 ml of a different buffer solution (measure the first buffer volume and then fill the other tubes to the same level), keeping the solutions in numerical order. Place two drops of methyl orange indicator in each tube and shake. Note the pH range over which the indicator changes colour, and estimate the pH at the middle of this range.

Repeat using methyl red indicator.

RESULTS

Working range of methyl orange = pH _____ to pH _____

pH at middle of range   = _____ = pK_{ia}

∴ Indicator constant K_{ia} = _____

Working range of methyl red   = pH_____ to pH_____

pH at middle of range   = _____ = pK_{ia}

∴ Indicator constant K_{ia} = _____

THEORY

In a solution of an acid HA there exists the equilibrium

$$HA \rightleftharpoons H^+ + A^-$$

and hence, by the laws of chemical equilibrium,

$$[H^+] = K_a \cdot \frac{[HA]}{[A^-]}$$

where K_a is the equilibrium constant, also known in this case as the acid dissociation constant. For a weak acid the equation may be written

$$[H^+] = K_a \frac{[\text{acid}]}{[\text{anion}]}$$

since dissociation is only slight and therefore [HA] and total acid concentration are almost equal.

If the solution contains an appreciable quantity of a salt of the acid as well as the acid itself, then the equation may be written

$$[H^+] = K_a \frac{[\text{acid}]}{[\text{salt}]}$$

since the highly dissociated salt may be regarded as providing practically all the anions and the salt also suppresses the dissociation of the acid by the common ion effect. It is convenient to express the relationship as

$$pH = pK_a - \log \frac{[\text{acid}]}{[\text{salt}]} \quad (\text{where } pX = -\log_{10} X)$$

Thus a solution of definite pH is produced by a solution of the acid and its salt (usually Na^+ or K^+) in a definite molar ratio. Such a solution is known as a BUFFER SOLUTION as its pH is hardly affected by the addition of small quantities of a free acid or base, or by reasonable dilution. The buffer solutions prepared in this experiment utilise the weak tribasic citric acid and its sodium salts. The pK_a values for each dissociation are 3·13, 4·76 and 6·40, and thus it is able to provide three series of buffer solutions in the ranges 1–4, 4–5·6 and 5·9–7.

A pH indicator is usually a weak acid or a weak base (or a salt of one of these) which, according to the Ostwald theory, changes colour as it dissociates.

Consider the equilibrium in an aqueous solution of bromothymol blue (a weak acid type of indicator):

$$HIn \rightleftharpoons In^- + H^+$$

or

$$\text{Yellow} \rightleftharpoons \text{Blue} + H^+$$

Then

$$[H^+] = K_{ia} \frac{[\text{Yellow}]}{[\text{Blue}]}$$

or

$$pH = pK_{ia} - \log \frac{[\text{Yellow}]}{[\text{Blue}]}$$

where K_{ia} is the indicator dissociation constant (in this case the indicator *acid* dissociation constant).

At the middle of the indicator's working range

$$[\text{Yellow}] = [\text{Blue}]$$

and so

$$\text{pH} = \text{p}K_{ia}$$

In this experiment methyl orange and methyl red are used, and these are both indicators of the weak base type, dissociating

$$\text{InOH} \rightleftharpoons \text{In}^+ + \text{OH}^-$$

or

$$\text{Yellow} \rightleftharpoons \text{Red} + \text{OH}^-$$

Thus

$$\text{pOH} = \text{p}K_{ib} - \log \frac{[\text{Yellow}]}{[\text{Red}]}$$

and K_{ib} is the indicator *basic* dissociation constant.

At the middle of the indicator's working range

$$[\text{Yellow}] = [\text{Red}]$$

and so

$$\text{pOH} = \text{p}K_{ib}$$

or

$$\text{pH} = \text{p}K_w - \text{pOH} = \text{p}K_w - \text{p}K_{ib}$$

Since it is convenient to refer all indicator dissociation constants (whether acidic or basic) to the pH scale, it is usual to attribute 'acid' dissociation constants to basic indicators by using the expression

$$\text{p}K_{ia} = \text{p}K_w - \text{p}K_{ib}$$

$$14 - \text{p}K_{ib} \quad \text{(at room temperature)}$$

EXPERIMENT 12. APPROXIMATE DETERMINATION OF ORGANIC ACID DISSOCIATION CONSTANTS USING BUFFER SOLUTIONS

The buffer solutions prepared in Experiment 11 are required for this experiment.

APPARATUS AND CHEMICALS REQUIRED

25 clean test-tubes of about the same bore, with sufficient racks to accommodate them; dropping pipette with rubber teat; three 250-ml graduated flasks; wash bottle; 2-ml graduated pipette; two 5-ml pipettes; safety pipette filler; 250-ml beaker; 10-ml measuring cylinder; stirring rod.

Citrate buffers from Experiment 11; glacial acetic acid (3 ml); formic acid (2 ml); monochloroacetic acid (5 g); 0·1M (0·1N) sodium hydroxide (25 ml); universal indicator B.T.L. (5 ml).

EXPERIMENTAL PROCEDURE

Place 10 ml (approx.) portions of the buffer solutions, in numerical order, in test-tubes in the rack. (Measure the first buffer volume and then fill the other tubes to the same level.) Add three drops of universal pH indicator to each tube and shake to mix. It is essential that the indicator concentration is about the same in each buffer solution. This set of indicator-treated buffer solutions provides a fairly reliable colour comparator for the indicator being used. Cork and reserve the colour set for use in this experiment and in Experiment 13.

Prepare the following acid solutions:

0·2M acetic acid (2·86 ml glacial acid made up to 250 ml)
0·2M formic acid (1·87 ml formic acid made up to 250 ml)
0·2M monochloroacetic acid (4·725 g dissolved and made up to 250 ml)

Pipette 5 ml of 0·2M acetic acid and 5 ml of 0·1M sodium hydroxide into a clean, dry test-tube (this produces half-neutralised acid), add three drops of universal indicator and shake. Now determine the approximate pH of this solution by comparison with the colour set, interpolating if necessary. Record the pH and then repeat using formic acid and monochloroacetic acid in turn.

RESULTS

Acid	pH for half-neutralised acid $pH = pK_a$	Acid dissociation constant $K_a = -\text{antilog } pK_a$
Acetic		
Formic		
Monochloroacetic		

THEORY

Read the theory of buffer solutions at the end of Experiment 11.

For a mixed solution of a weak acid and one of its salts the pH is given by

$$pH = pK_a - \log \frac{[acid]}{[salt]}$$

Thus $pH = pK_a$ when $[acid] = [salt]$, *i.e.* for half-neutralised acid.

A 'universal indicator' is a mixture of colour-change pH indicators chosen so that the colour change may be extended over a wide pH range. The colour changes are usually arranged for convenience in the order of the solar spectrum. It is hardly worth while preparing universal indicators as they are readily available commercially. Their sole use is in the determination of approximate pH.

EXPERIMENT 13. TO PLOT A TITRATION CURVE WITH THE AID OF A UNIVERSAL INDICATOR AND BUFFER SOLUTIONS

The reaction to be followed is one between a strong acid (hydrochloric acid) and a very weak base (pyridine). This experiment utilises the colour set prepared in Experiment 12.

APPARATUS AND CHEMICALS REQUIRED

Burette and stand; 50-ml pipette; pipette safety filler; 250-ml conical flask; dropping pipette with rubber teat; 100-ml measuring cylinder; 250-ml beaker; funnel; 2-ml pipette; 250-ml graduated flask; clean test-tube; 10-ml measuring cylinder.

Colour set from Experiment 12 (shake the tubes well before use); 0·1M (0·1N) hydrochloric acid (75 ml); pyridine (2 ml); universal indicator B.T.L. (4 ml).

EXPERIMENTAL PROCEDURE

Prepare approximately 0·1M (0·1N) solution of pyridine (2 ml of the base, using a safety pipette filler, made up to 250 ml with distilled water). This solution is best prepared in a fume cupboard as the vapour of undiluted pyridine is offensive. The dilute solution may be used in the open laboratory without discomfort.

Place approximately 4 ml of universal indicator in the flask of pyridine solution and shake well. Charge the burette with this indicator-treated solution. Pipette 50 ml of 0·1N hydrochloric acid into the conical flask and add 15 drops of universal indicator. (All solutions must contain universal indicator at the same concentration as the solutions in the colour set.)

Run 20 ml of pyridine solution into the acid in the conical flask, agitate, and determine the approximate pH of the solution by comparing it with the tubes of the colour set. (This comparison is best done by pouring a little of the mixture into a test-tube and using this same depth of liquid. Return the sample to the conical flask before proceeding with the titration.) Determine the approximate pH's after further additions of 10, 10, 5, 5, 5, 5, 5, 5, 5, 10 and 15 ml of base.

Tabulate results and plot pH against total volume of base added. Estimate the equivalence-point and calculate the normality of the indicator-treated pyridine solution.

RESULTS

Volume additions of pyridine solution (ml)	Total volume of pyridine solution added (ml)	pH
20	20	
10	30	
10	40	
etc.	etc.	

Equivalence-point (estimated from the curve obtained):

                50 ml 0·1N hydrochloric acid ≡ _____ ml pyridine solution

Normality of pyridine solution = _____ = _____N

THEORY

From the curve obtained in this experiment it will be seen that the pH change near to the equivalence-point is only about 2·5 units for an addition of about 15 ml of base. A single colour-change indicator could not, therefore, be expected to provide an accurate method of detecting the equivalence-point for such a titration (strong acid—very weak base). If electrical apparatus is not available, the use of a universal indicator and the plotting of a titration curve reduces the colorimetric inaccuracies to a minimum. An error of ±1% may be expected, compared with an error of about ±5% using the only suitable single colour change indicator, methyl orange.

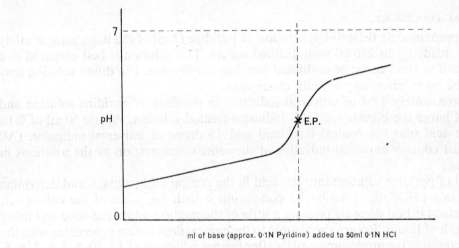

ml of base (approx. 0·1N Pyridine) added to 50ml 0·1N HCl

Fig. 13.1

Transition Temperatures

EXPERIMENT 14. TO DETERMINE THE TEMPERATURE OF TRANSITION BETWEEN TWO OF THE CRYSTALLINE FORMS OF AMMONIUM NITRATE BY A DILATOMETER METHOD (POLYMORPHIC TRANSITION TEMPERATURE)

APPARATUS AND CHEMICALS REQUIRED

1 metre length of capillary tubing (bore 2–3 mm); boiling tube with single bored rubber bung to fit well; metre rule; 500-ml beaker (preferably of the tall type); bunsen with barrel removed; three stands and clamps; beaker clamp; pestle and mortar; 0–110°C (in 1°C) thermometer; ring stirrer for water bath (the beaker).

Ammonium nitrate (30 g); liquid paraffin (oil bath grade) (75 ml).

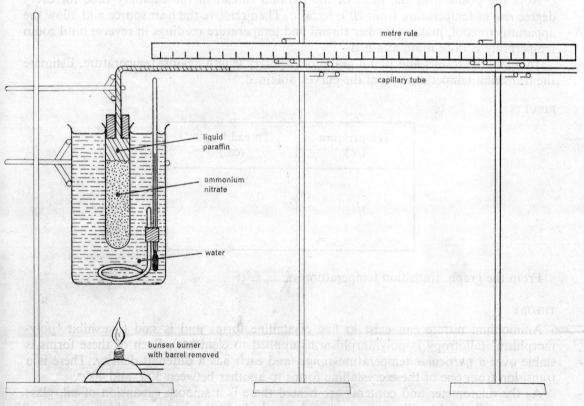

Fig. 14.1

EXPERIMENTAL PROCEDURE

The general arrangement of the apparatus is shown in the diagram but care must be taken in assembling it if good results are to be obtained. Bend the capillary tube to form a right angle about 10 cm from one end (take care to warm gently at first or the thick-walled tubing will crack). Moisten the hole in the bung and insert the short arm of the bent capillary tube till the end of it is flush with the lower surface of the bung.

Half fill the boiling tube with liquid paraffin and then add ammonium nitrate (the lumps should be gently crushed to an even crystal size and between 25 g and 30 g added) in small quantities, stirring gently to ensure the release of all air bubbles. Top up with liquid paraffin to fractionally above the level which will be reached by the bung when it is inserted. Carefully wipe the top inside of the boiling tube free from oil (otherwise the bung will slip out) and insert the moistened bung with the attached capillary tube. At this stage all air bubbles should have been removed from the tube and a thread of oil pushed into the capillary. Assemble the rest of the apparatus as shown, *away from draughts*. It is convenient to attach a ring stirrer to the thermometer with a small piece of rubber tubing.

Gently heat the water bath so that the temperature rises no more than a degree a minute. (This rate should be achieved by using the modified bunsen with a 1-cm flame about 4 cm beneath the bath.) Stir continuously.

Note the position of the head of the paraffin thread in the capillary tube for every degree rise of temperature from 20°c to 50°c. Then remove the heat source and allow the apparatus to cool, making further thread and temperature readings in reverse until room temperature is once more attained.

Tabulate the results and plot a graph of capillary length against temperature. Estimate the transition temperature from the curves obtained.

RESULTS

Temperature (°c)	Thread length (cm)

From the graph, transition temperature = _____°c

THEORY

Ammonium nitrate can exist in five crystalline forms and is said to exhibit 'polymorphism' (allotropy is polymorphism as applied to elements). Each of these forms is stable over a particular temperature range, and each has a different density. There is a transition from one of these crystalline forms to another between 30°c and 40°c.

As the dilatometer and contents are heated there is a smooth expansion of oil, glass and salt until the transition temperature is reached. At this point a sudden increase of volume occurs, as the salt changes to a new, less dense, crystal form. A similar, reverse change takes place on cooling. The curves obtained are of the type

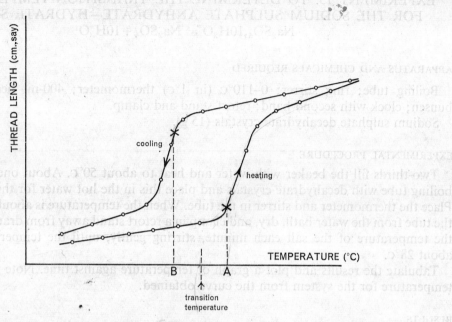

Fig. 14.2

There is always a time-lag in transitions in the solid state and this is why very slow heating and cooling are essential if transitions are to be observed satisfactorily. The transition temperature is given by the mean of A and B.

The same type of experiment may be used to determine the temperature of transition between rhombic and monoclinic sulphur, but considerable skill and patience are required to attain a constant and very slow temperature elevation of the heating bath between 90°c and 100°c—a temperature far removed from that of the laboratory. If this is attempted the dilatometer and contents must be heated to about 80°c before the capillary tube is attached. 30% sulphuric acid may be used in the dilatometer in place of liquid paraffin, and a 30% solution of calcium chloride in the heating bath enables temperatures a few degrees above 100°c to be attained. (It would be potentially dangerous to use an oil bath.)

EXPERIMENT 15. TO DETERMINE THE TRANSITION TEMPERATURE FOR THE SODIUM SULPHATE ANHYDRATE—HYDRATE SYSTEM
$$Na_2SO_4,10H_2O \rightleftharpoons Na_2SO_4 + 10H_2O$$

APPARATUS AND CHEMICALS REQUIRED

Boiling tube; ring stirrer; $0-110°c$ (in $1°c$) thermometer; 400-ml beaker; tripod; bunsen; clock with second hand; retort stand and clamp.

Sodium sulphate decahydrate crystals (15 g).

EXPERIMENTAL PROCEDURE

Two-thirds fill the beaker with water and heat to about $50°c$. About one-third fill the boiling tube with decahydrate crystals and place this in the hot water for the salt to melt. Place the thermometer and stirrer in the tube. When the temperature is about $40°c$ remove the tube from the water bath, dry, and clamp in a retort stand away from draughts. Record the temperature of the salt each minute, stirring gently, until the temperature falls to about $28°c$.

Tabulate the results and plot a graph of temperature against time. Note the transition temperature for the system from the curve obtained.

RESULTS

Temperature ($°c$)	Time (min)

Transition temperature for the system = _____$°c$

THEORY

Considering a given salt which may exist in the anhydrous and in hydrated forms, the following rules will generally apply to its formation of an aqueous solution:

(1) The heat of solution of the anhydrate is exothermic and that of the highest hydrate endothermic, intermediate hydrates being graded between the two extremes.

(2) Each hydrate is stable only within a definite temperature range.

(3) In accord with (1) and (2), the solubility curve will have distinct sections corresponding with the temperature ranges of hydrate stability, and the average slope of each section will be greater the higher the hydrate, perhaps becoming negative for lower hydrate and anhydrate forms (Figs. 15.1 and 15.2).

36

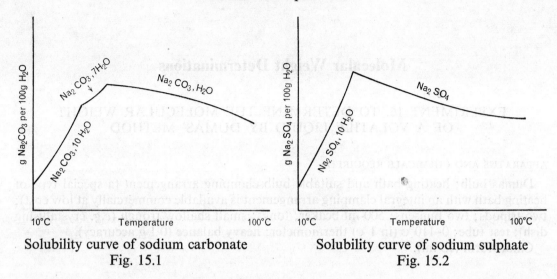

Solubility curve of sodium carbonate
Fig. 15.1

Solubility curve of sodium sulphate
Fig. 15.2

The discontinuities in the curves are called 'transition points' and the temperatures at which they occur 'transition temperatures'.

Upon heating, most hydrates lose some, if not all, of their water of crystallisation below 100°C, and the usual result is the formation of a solution; the salt melts into solution in its own discarded water of crystallisation, since the latter does not vaporise rapidly below 100°C. In the case of sodium sulphate, in this experiment, heating above the transition temperature of the decahydrate results in a slurry of anhydrate suspended in a saturated solution.

As the melt cools the saturated solution becomes more concentrated (note the negative slope of the anhydrate solubility curve, Fig. 15.2) until the transition temperature is reached. At this point the decahydrate becomes the stable form and begins to crystallise from the saturated solution. The crystallisation of the decahydrate (endothermic heat of solution) liberates heat and the cooling is arrested until crystallisation is almost complete. This cooling arrest at the transition temperature results in a horizontal section in the experimental cooling curve which is of the type shown below.

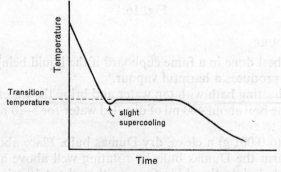

Fig. 15.3

Molecular Weight Determinations

EXPERIMENT 16. TO DETERMINE THE MOLECULAR WEIGHT OF A VOLATILE LIQUID BY DUMAS' METHOD

APPARATUS AND CHEMICALS REQUIRED

Dumas bulb; heating bath and suitable bulb clamping arrangement (a special type of heating bath with an integral clamping arrangement is available commercially at low cost); two tripods; two bunsens; 500-ml beaker; tongs; small shallow trough (*e.g.* crystallising dish); test tube; 0–110°C (in 1°C) thermometer; heavy balance (0·1 g accuracy).

Chloroform (15 ml).

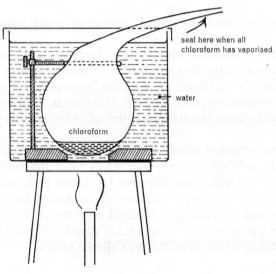

seal here when all
chloroform has vaporised

water

chloroform

Fig. 16.1

EXPERIMENTAL PROCEDURE

This experiment is best done in a fume cupboard if the liquid being investigated (in this instance chloroform) produces a harmful vapour.

Two-thirds fill the heating bath with tap water and bring the water almost to its boiling point. At the same time boil about 400 ml of distilled water for 5–10 min to expel dissolved air and put this aside to cool.

Weigh accurately (to 0·001 g) a clean, dry Dumas bulb. Place about 15 ml chloroform in a test-tube and warm the Dumas bulb by rotating well above a bunsen flame. Then insert the neck of the bulb into the chloroform so that about 10 ml of the liquid is drawn into the bulb as it cools. Clamp the bulb in the hot water bath, immersing as much of the bulb and neck as possible, and bring the water to the boil. Leave the bulb in the boiling water for about 20 min.

Read the barometer and record the balance room temperature.

With the bulb still immersed in boiling water, seal the neck by heating about an inch from the tip and drawing out. Retain the piece of glass drawn off. (Use the hottest bunsen flame and a pair of tongs for the sealing process.) Remove the bulb from the water bath, dry thoroughly, and then reweigh it, together with the drawn-off piece.

Transfer the air-free water to a small trough. Place the neck of the bulb under the water and break the tip by applying pressure with tongs. Water should then enter and fill the bulb to within at least 2 ml (if much air is found to be left in the bulb the experiment must be repeated). Weigh the water-filled bulb, together with any large pieces of glass, to within 0·1 g on a rough balance.

Calculate the molecular weight of chloroform and the probable error in the result. Express the result to an appropriate order of accuracy.

RESULTS

Weight of bulb in air	$= a$ g
Weight of bulb and chloroform vapour	$= b$ g
Weight of bulb filled with water	$= c$ g
Barometric pressure	$= P$ mm Hg
Balance room temperature	$= T°$c
Heating bath temperature	$= 100°$c
Density of air at 0°c and 760 mm Hg	$= 1·29$ g/l
Density of hydrogen at 0°c and 760 mm Hg	$= 0·089$ g/l

CALCULATION

Weight of water in bulb $= (c-a)$ g

Volume of bulb $= (c-a)$ ml

Volume of air in bulb reduced to 0°c and 760 mm
$$= (c-a) \cdot \frac{273}{273+T} \cdot \frac{P}{760} \text{ ml}$$

Weight of air in bulb
$$= (c-a) \cdot \frac{273}{273+T} \cdot \frac{P}{760} \cdot \frac{1·29}{1000} \text{ g}$$
$$= d \text{ g}$$

Weight of vacuous bulb $= (a-d)$ g

Weight of chloroform vapour in bulb $= b-(a-d)$ g $= e$ g

Volume that chloroform vapour would occupy (if that were possible) at 0°c and 760 mm
$$= (c-a) \cdot \frac{273}{373} \cdot \frac{P}{760} \text{ ml}$$

Weight of an equal volume of hydrogen $= (c-a) \cdot \dfrac{273}{373} \cdot \dfrac{P}{760} \cdot \dfrac{0 \cdot 089}{1000}$ ml

$\qquad\qquad\qquad\qquad\qquad\qquad\qquad = f$ g

Vapour density of chloroform $= \dfrac{\text{Weight of chloroform vapour}}{\text{Weight of an equal volume of hydrogen at same temperature and pressure}}$

$\qquad\qquad\qquad\qquad\qquad\qquad = \dfrac{e}{f}$

Molecular weight of chloroform $= \dfrac{2e}{f}$

NOTE

Griffin & George Ltd advertise a modified Dumas bulb. This modification is essentially the provision of an accurately-ground glass closure cap which eliminates the need for sealing. The volatile liquid is injected into the bulb with a hypodermic syringe.

EXPERIMENT 17. TO DETERMINE THE MOLECULAR WEIGHT OF A VOLATILE LIQUID BY THE VICTOR MEYER METHOD

APPARATUS AND CHEMICALS REQUIRED

Victor Meyer apparatus, preferably fitted with a steam jacket of the type shown in the diagram; tapless burette with attached levelling reservoir as shown; stands with clamps and funnel ring; steam can; tripod; bunsen; Hofmann bottle (and half-bored cork for carrying); 0–110°C (in 1°C) thermometer; rubber and glass tubing and rubber bungs as shown; dropping pipette with rubber teat.

Acetone (1 ml); mercury (2-3 ml).

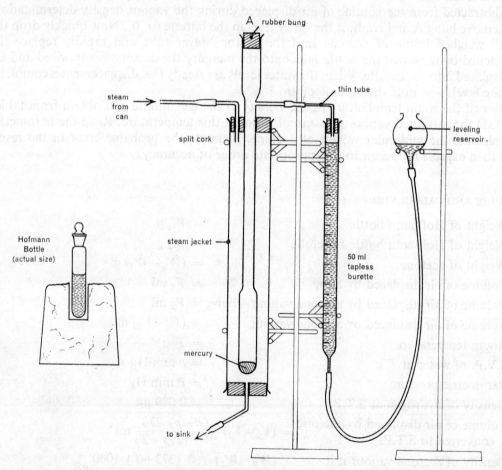

Fig. 17.1

41

EXPERIMENTAL PROCEDURE

The modified apparatus is easier to use than the conventional type and is capable of giving better results. It is particularly easy to measure the volume of air displaced.

With the apparatus assembled remove bung A (and leave out), adjust the water level in the burette to the '0' mark, and pour a little mercury (1 cm depth) into the Victor Meyer tube. Pass steam through the vapour jacket and leave, with steam flowing, to attain thermal equilibrium.

Weigh accurately (to 0·001 g) a clean, dry Hofmann bottle complete with well-fitting ground glass stopper. Half to two-thirds fill the bottle with acetone with the aid of a dropping pipette, replace the stopper, dry the outside, then reweigh. Carry the weighed bottle and contents in the vertical position in the half-bored cork.

When steam has been passing for about 15 min, moisten the bung A and quickly insert it at the top of the Victor Meyer tube. Some air will be displaced into the burette. Equate levels and note the volume of air displaced. This bung displacement volume must be subtracted from the volume of air displaced during the vapour density determination.

Remove bung A and readjust the water level in the burette to '0'. Now quickly drop the small weighed bottle of acetone into the Victor Meyer tube and rapidly replace the moistened bung. When the bottle falls onto the mercury the acetone is vaporised and air is displaced into the burette. When the water levels are steady (*i.e.* displacement is complete) equate levels and note the volume of air.

Record the room temperature in the vicinity of the gas burette and obtain from tables (p. 131) the saturated vapour pressure of water at this temperature. Read the barometer.

Calculate the molecular weight of acetone. Estimate the probable error in the result and then express the result to an appropriate order of accuracy.

RESULTS AND CALCULATION

Weight of Hofmann bottle	$= W_1$ g
Weight of Hofmann bottle + acetone	$= W_2$ g
Weight of acetone	$= (W_2 - W_1)$ g
Volume of air displaced by bung	$= V_1$ ml
Volume of air displaced by acetone vapour + bung	$= V_2$ ml
Volume of air displaced by acetone vapour	$= (V_2 - V_1)$ ml
Room temperature	$= T°$c
S.V.P. of water at $T°$c	$= p$ mm Hg
Barometric pressure	$= P$ mm Hg
Density of hydrogen at S.T.P.	$= 0·089$ g/l

$$\text{Volume of air displaced by acetone converted to S.T.P.} = (V_2 - V_1) \frac{(P-p)\ 273}{760\ (273+T)}\ \text{ml}$$

$$\text{Density of acetone vapour if it could exist at S.T.P.} = \frac{(W_2 - W_1)\ 760\ (273+T)\ 1000}{(V_2 - V_1)\ (P-p)\ 273}\ \text{g/l}$$

Vapour density of acetone $= \dfrac{\text{Density of acetone vapour}}{\substack{\text{Density hydrogen at same}\\\text{temperature and pressure}}}$

$$= \frac{(W_2 - W_1)\ 760\ (273+T)\ 1000}{(V_2 - V_1)\ (P-p)\ 273 \times 0.089}$$

$$= d$$

Molecular weight of acetone $= 2d$

THEORY

As the acetone vaporises it expands within the bulb of the Victor Meyer tube and forces air out (it is essential that no acetone vapour leaves the apparatus). The air contracts and the measured volume represents the volume which the acetone vapour itself would occupy at room temperature if it could do so without condensing. Converting the volume to s.t.p. the (hypothetical) density of the acetone vapour under these conditions may be calculated and compared with the density of hydrogen to give the relative (vapour) density and hence the molecular weight.

NOTE

If a gas syringe of suitable volume is available, this may be used to collect the displaced air. The syringe must of course be clamped in the horizontal position. If a syringe is used there is no need to make the correction for the s.v.p. of water.

EXPERIMENT 18. TO DETERMINE THE DENSITY (and hence MOLECULAR WEIGHT) OF A GAS BY A DIRECT METHOD

APPARATUS AND CHEMICALS REQUIRED

250-ml flat-bottomed flask fitted with bung, tubes and pinch clips as shown in the diagram; 500-ml measuring cylinder; two dreschel bottles; rubber and glass tubing of appropriate diameters; 0–110°c (in 1°c) thermometer.

Cylinder of carbon dioxide (or a Kipp's apparatus charged for carbon dioxide and fitted to wash the gas first with water and then with conc. sulphuric acid); conc. sulphuric acid.

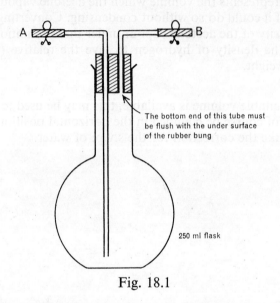

The bottom end of this tube must be flush with the under surface of the rubber bung

250 ml flask

Fig. 18.1

EXPERIMENTAL PROCEDURE

Weigh the clean dry flask (complete with tubes and clips) to within 0·001 g.

With both clips open connect A to a source of dry carbon dioxide (bubble carefully through conc. sulphuric acid) and allow the gas to flow slowly through the flask for 15 min. Turn off the gas supply and then close clip A and then clip B. Reweigh the flask (again to 0·001 g).

Open both clips and connect A to a water tap. Completely fill the flask and tubes with water, close the clips, and remove from the water supply. Drain the water from the flask (via B) into a 500-ml measuring cylinder and note the volume of the water.

Record the barometric pressure and the room temperature.

Calculate the density (g/l) of carbon dioxide and hence its molecular weight. Estimate the probable error in the result and then express the result to an appropriate order of accuracy.

The experiment may be repeated with other gases (*e.g.* sulphur dioxide, butane) if convenient sources of them are available.

RESULTS

Weight of flask containing air	$= W_1$ g
Weight of flask containing carbon dioxide	$= W_2$ g
Volume of water contained by flask	$= V$ ml
Barometric pressure	$= P$ mm Hg
Room temperature	$= T$°C
Density of air (0°C and 760 mm)	$= 1\cdot29$ g/l
Density of hydrogen (0°C and 760 mm)	$= 0\cdot089$ g/l

CALCULATION

Volume of air in flask reduced to S.T.P. $= \dfrac{273\ P\ V}{(273+T)\ 760}$ ml

Weight of air in flask

$$= \dfrac{273\ P\ V\ 1\cdot29}{(273+T)760 \times 1000}\ \text{g}$$

$$= W\ \text{g}$$

Weight of vacuous flask $= (W_1 - W)$ g

Weight of carbon dioxide filling flask $= W_2 - (W_1 - W)$ g

Density of carbon dioxide at S.T.P. $= \dfrac{[W_2 - (W_1 - W)]\ 1000(273+T)\ 760}{273\ P\ V}$ g/l

$$= D\ \text{g/l}$$

Vapour density (relative density) $= \dfrac{\text{Density of carbon dioxide}}{\text{Density of hydrogen}}$ at same temp. and pres.

$$= \dfrac{D}{0\cdot089}$$

Molecular weight $= \dfrac{2D}{0\cdot089} = \underline{\hspace{3cm}}$

EXPERIMENT 19. TO DETERMINE THE MOLECULAR WEIGHT OF CHLOROBENZENE BY STEAM DISTILLATION

One of the distillation assemblies shown, using a 500-ml flask and a 50–105°C (in 0·1°C) thermometer; two 250-ml conical flasks; 250-ml separating funnel and suitable funnel ring and stand; two 100-ml (in 1 ml) measuring cylinders; magnifying lens.
Chlorobenzene (150 ml).

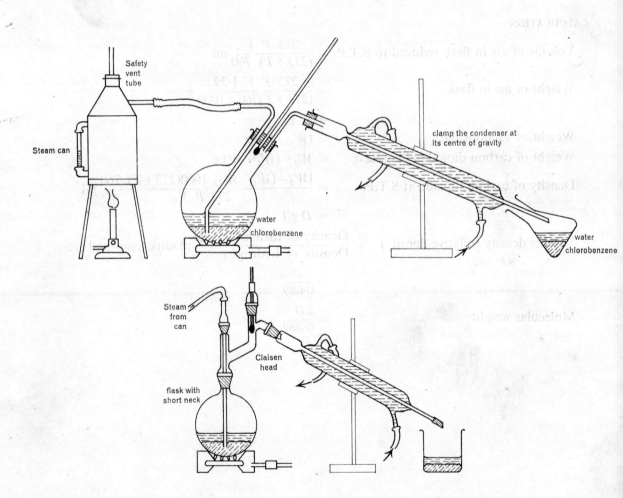

Fig. 19.1

46

EXPERIMENTAL PROCEDURE

(1) Calibration of the thermometer

The thermometer must first be calibrated at about the temperature at which it is to be used. (This is an essential preliminary since a fraction of a degree error in measuring the final steam distillation temperature introduces a large error in the water s.v.p. value obtained from tables.) Place about 200 ml of tap water in the distillation flask, heat almost to boiling and then pass in steam from the can. When the water is distilling steadily read the thermometer with the aid of a lens and note the steady temperature (to 0·1°c). Read the barometer and use the graph on page 132 to find the true boiling point of water at the noted pressure. Calculate the thermometer correction.

(2) Determination of molecular weight

Empty the flask and recharge it with about 150 ml of chlorobenzene and 75 ml of water. Warm this mixture and pass in steam. When distillation is proceeding steadily discard the first 20 ml (approx.) of distillate and then collect a fraction of about 100 ml in a clean conical flask. Note the steady distillation temperature (to 0·1°c).

Separate the two layers of the distillate (using a separating funnel), running each into a clean 100-ml measuring cylinder. Note the volumes. (The layers are separated before measuring their volumes because the curvature of the interface is considerable and would lead to unnecessary error.) Use the graph on p. 132 to find the s.v.p. of water at the (corrected) distillation temperature.

Calculate the molecular weight of chlorobenzene. Estimate the probable error in the result and then express the result to an appropriate order of accuracy.

RESULTS AND CALCULATION

Barometric pressure $= P$ mm Hg

Thermometer reading during water distillation $= T_1°c$

B.P. water at P mm Hg (from tables) $= T_2°c$

Thermometer correction $= (T_2 - T_1)$ $= \Delta T°c$

Volume of chlorobenzene in distilled fraction $= V_c$ ml

Volume (and weight) of water in distilled fraction $= V_w$ ml (g)

Density of chlorobenzene (at 20°C) $= 1·11$ g/ml

Weight of chlorobenzene in distilled fraction $= 1·11 . V_c$ g

Thermometer reading during mixture distillation $= T_3°c$

Corrected distillation temperature $= (T_3 + \Delta T)°c$

S.V.P. water at $(T_3 + \Delta T)°c$ (from tables) $= p_w$ mm Hg

Then M.W. chlorobenzene M_c is given by

$$\frac{1·11 \times V_c}{V_w} = \frac{M_c}{18} \times \frac{P - p_w}{p_w}$$

THEORY

A liquid boils when its saturated vapour pressure becomes equal to its environmental pressure (in this case atmospheric pressure).

For a mixture of immiscible liquids, as in steam distillation, each liquid exerts the vapour pressure it would exert if the other were absent. Thus the vapour pressure of the mixture is independent of composition and is given by

$$P_m = p_A + p_B$$

where P_m is the vapour pressure of the mixture, and p_A and p_B are the vapour pressures of the pure liquids A and B.

Also since P_m is greater than either p_A or p_B the mixture will boil at a lower temperature than either pure A or pure B.

Now

$$\frac{\text{Weight of chlorobenzene in the distillate}}{\text{Weight of water in the distillate}}$$

$$= \frac{\text{Weight of chlorobenzene in vapour}}{\text{Weight of steam in vapour}}$$

$$= \frac{\text{Density of chlorobenzene vapour}}{\text{Density of steam}} \times \frac{\text{volume of chlorobenzene vapour}}{\text{volume of steam}}$$

$$= \frac{\text{M.W. of chlorobenzene}}{\text{M.W. of steam}} \times \frac{\text{partial pressure of chlorobenzene}}{\text{partial pressure of steam}}$$

$$= \frac{\text{M.W. of chlorobenzene}}{\text{M.W. of steam}} \times \frac{\text{Atmos. pressure} - \text{partial pressure of steam}}{\text{partial pressure of steam}}$$

i.e.

$$\frac{1 \cdot 11 \times V_c}{V_w} = \frac{M_c}{18} \times \frac{P - p_w}{p_w}$$

EXPERIMENT 20. TO DETERMINE THE MOLECULAR WEIGHT OF UREA BY LANDSBERGER'S ELEVATION OF BOILING POINT METHOD

APPARATUS AND CHEMICALS REQUIRED

Landsberger's apparatus fitted with a 50–110°c (in 0·1°c thermometer); 250-ml conical flask; tripod; bunsen; magnifying lens; retort stand and clamps; bungs and glass and rubber tubing as shown; 10-ml measuring cylinder; T-piece; spring clip.

Urea (8 g); pumice (granular) or small pieces of broken pot.

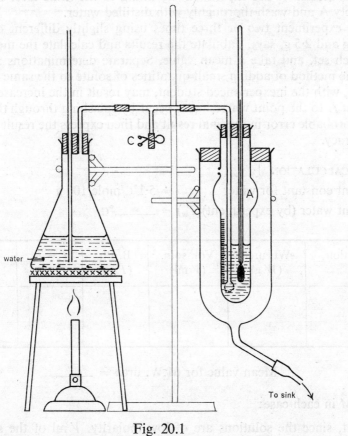

water

C

A

To sink

Fig. 20.1

EXPERIMENTAL PROCEDURE

Assemble the apparatus as shown in Fig. 20.1 and place about 150 ml distilled water and a little pumice (or porous pot) in the conical flask and about 5 ml distilled water in the graduated inner tube A. The vapour supply tube and the thermometer in A must be arranged so that the vapour does not impinge directly on to the thermometer bulb.

49

Weigh accurately (to 0·001 g) about 2 g of dry urea and put to one side.

Open the clip C and bring the water in the flask to the boil. Close C and boil the water steadily and gently so that a stream of vapour passes into the water in A, bringing the latter to the boil. Note the steady boiling temperature (T_w) to the nearest 0·02°C with the aid of a magnifying lens, and then immediately stop the flow of vapour to A by opening the clip C. Raise the thermometer and vapour tube from A and add the weighed quantity of urea to the hot water, making sure that all of the added urea goes into solution. Replace the thermometer and vapour tube and close C. The passage of vapour brings the solution to the boil and the steady, elevated, boiling point (T_s) is noted. Open C, raise the thermometer and vapour tube, and note the volume of the solution. (Some examples of this apparatus are not graduated in ml, and in such cases the graduation is noted and the volume to this point is determined after the experiment by filling to the same level with water from a burette). Empty A and wash thoroughly with distilled water.

Repeat the experiment two or three times using slightly different quantities of urea (between 1·5 g and 2·5 g, say). Tabulate the results and calculate the molecular weight of urea from each set, and take a mean value. Separate determinations are recommended rather than the method of adding small quantities of solute to the same solution since the latter method, with the inexperienced student, may result in the increase in the volume of the solution in A to the point where losses occur by splashing through the small hole in A. Estimate the probable error in the final result and then express the result to an appropriate order of accuracy.

RESULTS AND CALCULATION

Boiling point constant for water   = 5·1°C/mole/100 g

Boiling point water (by experiment)(T_w) = _____°C

Determination No.	Wt. urea (W g)	Vol. soln. (V ml)	B. Pt. (T_s)°C	$T_s - T_w$	Mol. Wt. of urea(M)
1					
2					
etc.					

Mean value for M.W. urea = _____

To calculate M in each case:

Assume that, since the solutions are of low molarity, V ml of the solution contains V ml, and therefore $V \times 0.958$ g, of water. (S.G. water at 100°C = 0·958.)

$(T_s - T_w)$°C is elevation in $0.958V$ g water by W g urea

$\therefore$ 5·1°C is elevation in   100 g   water by $W \times \dfrac{5 \cdot 1}{(T_s - T_w)} \times \dfrac{100}{0 \cdot 958 V}$ g

$$= \text{M.W. urea}$$

EXPERIMENT 21. TO INVESTIGATE THE MOLECULAR STATE OF PHENYL-ACETIC ACID IN BENZENE SOLUTION BY A FREEZING POINT METHOD.

APPARATUS AND CHEMICALS REQUIRED

Beckmann freezing point apparatus (Fig. 21.1); −5 to 50°c (in 0·1°c) or, better, a benzene F.P. thermometer, −1 to 10°c (in 0·01°c); 25-ml pipette; pipette safety filler; magnifying lens.

Benzene (pure) (25 ml); phenylacetic acid (2·5 g); ice.

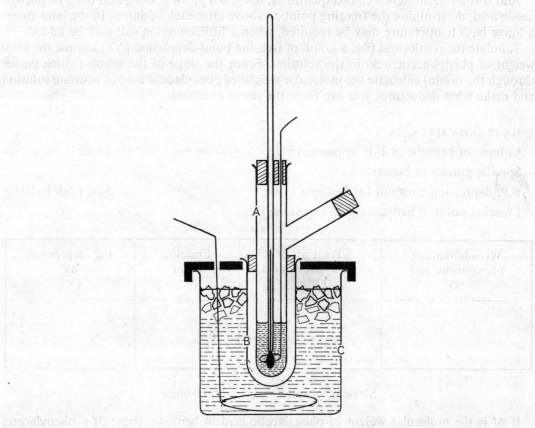

Fig. 21.1

EXPERIMENTAL PROCEDURE

Fill the unoccupied space of beaker C with a mixture of ice and water (use approximately equal bulks). Pipette 25 ml of benzene into the inner tube A using a safety filler.

N.B. Benzene is carcinogenic, and every care must be taken to avoid inhaling the vapour or spilling the liquid on the skin.

Allow the benzene to cool until its temperature is about 5°C. Stir vigorously and when crystals appear record the steady temperature to the nearest 0·02°C with the aid of a magnifying lens (the temperature may rise slightly upon crystallisation due to previous supercooling).

Weigh accurately about 0·4 g of phenylacetic acid (all weighings in this experiment to be within 0·001 g). Remove the inner tube A and allow the benzene to melt completely. Add the weighed quantity of phenylacetic acid to the benzene, taking care that all of the solid enters the liquid, and stir until all the acid has dissolved. Replace the tube in B and determine the freezing point of this solution, *i.e.* the temperature at which crystals *first* appear.

vital

Add further accurately weighed quantities, 0·4 g, 0·4 g, 0·4 g, 0·4 g and 0·4 g of phenylacetic acid, determining the freezing point as above after each addition. In the later stages a lower bath temperature may be required, when a little common salt may be added.

Tabulate the results and plot a graph of freezing point depression (ΔT) against the total weight of phenyl acetic acid in the solution. From the slope of the graph (which passes through the origin) calculate the molecular weight of phenylacetic acid in benzene solution and make what deductions you can from the result obtained.

RESULTS AND CALCULATION

Volume of benzene in F.P. apparatus	= 25 ml
Relative density ~~Specific gravity~~ of benzene	= 0·88
F.P. depression constant for benzene	= 51·2°C/mole/100 g
Freezing point of benzene (from experiment)	= _____°C

Wt. additions of phenylacetic acid (g)	Total wt. of phenylacetic acid (*W* g)	Freezing point (°C)	F.P. depression ΔT (°C)

Slope of graph = _____°C/g solute

If M is the molecular weight of phenylacetic acid in benzene, then: M g phenylacetic acid in 100 g benzene gives F.P. depression equal to 51·2°C.

Therefore, 1 g phenylacetic acid in 25 × 0·88 g benzene gives F.P. depression equal to

$$\frac{51·2 \times 100}{M \times 25 \times 0·88} °C = \text{slope of graph}$$

$$\therefore M = \frac{51·2 \times 100}{25 \times 0·88 \times \text{slope}} = \text{_____}$$

Molecular Weight Determinations (application)

EXPERIMENT 22. TO INVESTIGATE THE FORMATION OF THE MERCURI–IODIDE COMPLEX ION BY FREEZING POINT MEASUREMENTS

APPARATUS AND CHEMICALS REQUIRED

Freezing point apparatus (Fig. 21.1) fitted with a -5 to $50°c$ in $0.1°c$ thermometer; 25-ml pipette; pipette safety filler; magnifying lens.

$1.0M$ ($1.0N$) potassium iodide (30 ml); mercuric iodide (6·25 g); ice; saturated brine.

EXPERIMENTAL PROCEDURE

Charge the unoccupied space of the cooling bath B (Fig. 21.1) with a freezing mixture consisting of equal bulks of crushed ice and saturated brine. This mixture should be stirred frequently during the course of the experiment.

Pipette 25 ml of $1.0M$ potassium iodide into the inner tube A and place the thermometer and stirrer in position. Stir the solution vigorously as its temperature falls, and note its freezing point (the temperature at which crystals *first* appear) to the nearest $0.05°c$ with the aid of a magnifying lens. Vigorous stirring is essential since there will probably be a considerable tendency for supercooling to occur.

Weigh out about 1·25 g (to 0·01 g) of mercuric iodide. Remove the inner tube A from the assembly and allow the solution to warm up to about $0°c$. Add the mercuric iodide to the potassium iodide solution and stir until it has all dissolved. Determine the freezing point of this mercuri-iodide solution as before. Repeat four times by the further addition of similar quantities of mercuric iodide. The last portion will not all dissolve (the solution having become saturated with mercuric iodide) but the residue will not interfere with the last freezing point determination.

Tabulate the results and plot the freezing points of the solutions against moles of mercuric iodide dissolved (Fig. 22.1). The last freezing point cannot be plotted in the normal way since the weight of the mercuric iodide residue is unknown. However, knowing the freezing point of the saturated solution, the moles of mercuric iodide required for saturation may be read from the extrapolated curve.

RESULTS

Volume of $1.0M$ potassium iodide in A = 25 ml

Moles of potassium iodide in A   = 25/1000 = 0·025

Molecular weight of mercuric iodide   = 454

Wt. additions of mercuric iodide (g)	Total wt. of mercuric iodide in soln. (g)	Total moles of mercuric iodide in soln.	Freezing point of soln. (°C)
0	0	0	
	saturated	(value from graph)	

State your conclusions about the nature of the mercuri-iodide complex.

THEORY

Mercuric iodide is practically insoluble in water but dissolves in solutions of iodides to give complex anions. Possible reactions involved are

$$K^+ + I^- + HgI_2 \rightarrow K^+ + HgI_3^-$$ (I) (2 ions → 2 ions)
$$2K^+ + 2I^- + HgI_2 \rightarrow 2K^+ + HgI_4^{2-}$$ (II) (4 ions → 3 ions)
$$3K^+ + 3I^- + HgI_2 \rightarrow 3K^+ + HgI_5^{3-}$$ (III) (6 ions → 4 ions)

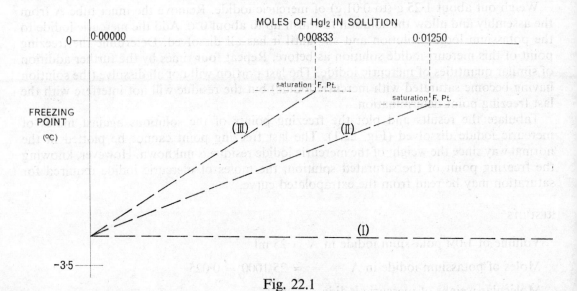

Fig. 22.1

In case (I) the number of particles present in solution would not change with the addition of mercuric iodide and the freezing point should remain approximately constant to give curve (I) (Fig. 22.1). In cases (II) and (III) the number of particles would steadily decrease and the freezing points should rise to give curves (II) and (III).

With 0·025 moles potassium iodide present curve (II) should terminate at the saturation point after the addition of $\frac{1}{2} \times 0·025$ moles of mercuric iodide, and curve (III) after the addition of $1/3 \times 0·025$ moles of mercuric iodide.

Phase Equilibria

EXPERIMENT 23. INVESTIGATION OF A BINARY SYSTEM PRODUCING A SIMPLE EUTECTIC

APPARATUS AND CHEMICALS REQUIRED

Two boiling tubes; ring stirrer; 400-ml beaker; tripod; bunsen; 0–110°c (in 1°c) thermometer; retort stand and clamps.

Naphthalene (12 g); *p*-nitrotoluene (14 g).

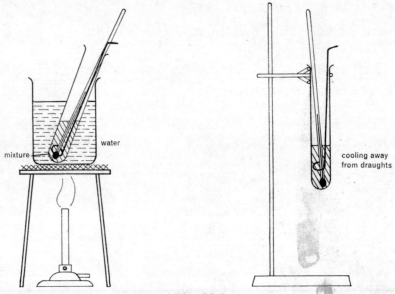

Fig. 23.1

EXPERIMENTAL PROCEDURE

Two-thirds fill the beaker with tap water and heat nearly to boiling.

Weigh 7·00 g of naphthalene into a boiling tube (all weighings in this experiment need only be within 0·01 g). Place the stirrer and thermometer in the boiling tube and then melt the naphthalene by immersion in the hot water. Remove the tube from the water bath, dry the outside, and position away from draughts. Stir continuously and record the temperature at which crystals first separate (super-cooling may occur and, if so, the higher, steady temperature should be recorded).

Weigh out 1·50 g of *p*-nitrotoluene, add this to the naphthalene in the boiling tube and repeat the above procedure (take care to mix the molten components thoroughly before removing the tube from the water bath). Add further quantities of 1·50, 1·50 and 2·50 g of *p*-nitrotoluene and determine the initial crystallisation temperatures in each case.

56

Repeat the whole of the above procedure, but starting with 7·00 g of *p*-nitrotoluene in a clean, dry boiling tube and adding 1·00, 1·00, 1·00 and 1·50 g quantities of naphthalene.

Tabulate the results, calculate the weight percentage of naphthalene in each mixture, and plot a temperature-composition curve for the system. Label the areas delineated by the curve and record the eutectic temperature and eutectic composition.

RESULTS

	1	2	3	4	5
Wt. naphthalene (g)	7·00	7·00	7·00	7·00	7·00
Wt. *p*-nitrotoluene added (g)	0	1·50	1·50	1·50	2·50
Total wt. *p*-nitrotoluene (g)	0	1·50	3·00	4·50	7·00
% naphthalene					
Initial cryst. temp. (°C)					

	6	7	8	9	10
Wt. *p*-nitrotoluene (g)	7·00	7·00	7·00	7·00	7·00
Wt. naphthalene added (g)	0	1·00	1·00	1·00	1·50
Total wt. naphthalene (g)	0	1·00	2·00	3·00	4·50
% naphthalene					
Initial cryst. temp. (°C)					

Eutectic temperature = _____°C

Eutectic composition = _____% *p*-nitrotoluene; _____% naphthalene

THEORY

Naphthalene and *p*-nitrotoluene neither combine to form a compound nor form a solid solution with one another.

The phase diagram is shown in Fig. 23.2 where A is the melting point of pure nitrotoluene. On adding naphthalene the freezing point of the mixture is lower than that of pure *p*-nitrotoluene. This is shown by AE which represents the compositions of solutions in equilibrium with solid *p*-nitrotoluene at different temperatures. Similarly BE shows the effect of adding naphthalene to *p*-nitrotoluene when once more the freezing points of different liquid mixtures, this time in equilibrium with solid naphthalene, are lowered along this curve.

At *E*, the intersection of AE and BE, both solid components can exist in equilibrium with a liquid of composition *E'* at a temperature *T'*. *E* is the 'eutectic point' and gives

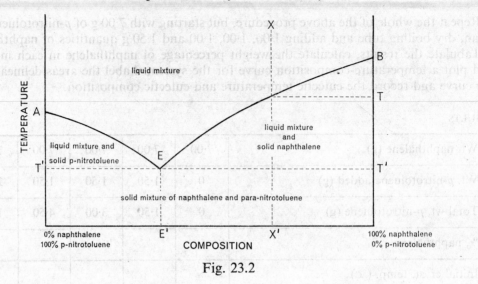

Fig. 23.2

the temperature and composition of the system when both components crystallise out simultaneously. This solid mixture is called the eutectic mixture. It has a definite fixed composition (at constant pressure) and melts and crystallises, as though it were a pure substance, at a fixed temperature.

If a liquid mixture of temperature and composition represented by X is cooled naphthalene crystals begin to separate at a temperature T from a liquid of composition X' and it is this temperature which is determined for a mixture of known composition in this experiment. This is repeated with several mixtures of known composition and thus points are obtained from which the curves AE and BE may be constructed.

EXPERIMENT 24. INVESTIGATION OF THE PHENOL-WATER SYSTEM

APPARATUS AND CHEMICALS REQUIRED

Eight clean, dry, corked boiling tubes and rack; 250-ml beaker; ring stirrer; 0–110°C (in 1°C) thermometer; bunsen; retort stand and clamps (including beaker clamp); burette and stand.

Phenol crystals (62 g).

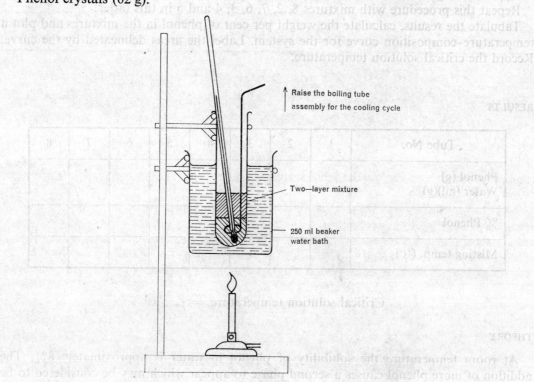

Raise the boiling tube
assembly for the cooling cycle

Two—layer mixture

250 ml beaker
water bath

Fig. 24.1

EXPERIMENTAL PROCEDURE

Label eight boiling tubes (near the top) and prepare in them the following phenol–water mixtures. Weigh the phenol crystals (to 0·01 g) into the boiling tubes direct and then run in the water from a burette. Cork the tubes.

Tube No.	1	2	3	4	5	6	7	8
Phenol (g)	2	2	4	6	10	11	13	14
Water (ml)(g)	23	18	16	14	15	9	7	6

Shake the mixtures well and then, on standing, each will be observed to separate into two layers. Phenol is caustic, and any contamination of the skin by the prepared mixtures must be removed immediately with soapy water.

Clamp mixture No. 1 in position in the water bath, insert the stirrer and thermometer, and heat the water bath gently while stirring the mixture briskly. Stop heating as soon as the phenol and water become completely miscible, *i.e.* when the mixture becomes clear. Remove the tube from the water bath, dry the outside, and position away from draughts.

Stir continuously and record the temperature at which the solution suddenly becomes misty, indicating the re-formation of two liquid phases.

Repeat this procedure with mixtures 8, 2, 7, 6, 3, 4 and 5 in this order.

Tabulate the results, calculate the weight per cent of phenol in the mixtures and plot a temperature–composition curve for the system. Label the areas delineated by the curve. Record the critical solution temperature.

RESULTS

Tube No.	1	2	3	4	5	6	7	8
Phenol (g) Water (ml)(g)								
% Phenol								
Misting temp. (°C)								

Critical solution temperature = ____°C

THEORY

At room temperature the solubility of phenol in water is approximately 8%. The addition of more phenol causes a second phase to appear which may be considered to be a solution of water in phenol. Now the solubility of phenol in water increases with increasing temperature and so at higher temperatures more phenol must be added to produce two liquid phases. Thus AB is the solubility curve of phenol in water. Similarly the solubility of water in phenol increases with increasing temperature as shown by CB.

The two curves meet at a certain temperature—known as the 'critical solution temperature'. Above this temperature two phases cannot exist, all water–phenol mixtures being completely miscible. The critical solution temperature is extremely sensitive to the presence of impurities and its determination is used for testing the purity of liquids.

When a homogeneous mixture of composition and temperature represented by P is cooled it becomes misty at a temperature T', signifying the formation of two phases of compositions X' and X''. The temperature at which the two layers reappear can generally be more accurately observed than that at which they merge.

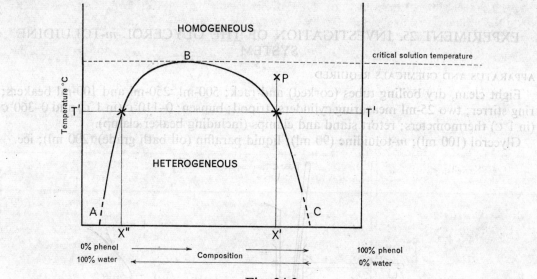

Fig. 24.2

In plotting temperature–composition curves for systems of this type, the composition may be expressed in terms of molar, weight, or volume fraction. Similar curves are obtained in each case. If further work is to be based on the curve obtained then it is probably best to express composition as a mole fraction.

EXPERIMENT 25. INVESTIGATION OF THE GLYCEROL–*m*-TOLUIDINE SYSTEM

APPARATUS AND CHEMICALS REQUIRED

Eight clean, dry boiling tubes (corked) and rack; 500-ml, 250-ml and 100-ml beakers; ring stirrer; two 25-ml measuring cylinders; tripod; bunsen; 0–110°c (in 1°c) and 0–360°c (in 1°c) thermometers; retort stand and clamps (including beaker clamp).

Glycerol (100 ml); *m*-toluidine (90 ml); liquid paraffin (oil bath grade) (200 ml); ice.

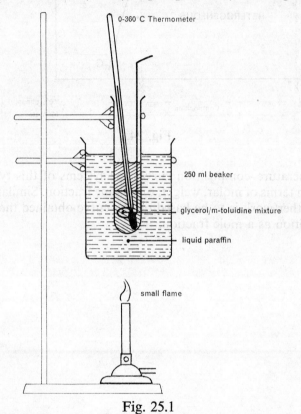

0-360 C Thermometer

250 ml beaker

glycerol/m-toluidine mixture

liquid paraffin

small flame

Fig. 25.1

EXPERIMENTAL PROCEDURE

Label eight boiling tubes (near the top) and prepare in them the following mixtures. Then cork the tubes.

Tube No.	1	2	3	4	5	6	7	8
m-Toluidine (ml)	4	4	6	10	11	14	13	20
Glycerol (ml)	19	16	12	10	6	5	3	3

Use measuring cylinders to obtain the required volumes. The glycerol should first be warmed to 40 or 50°C to reduce its viscosity and render it easier to handle. Should the skin be contaminated with *m*-toluidine, such contamination should be removed immediately with warm, soapy water.

Clamp tube No. 4 in position in the oil bath (see Fig. 25.1). Insert the stirrer and 0–360°C thermometer and gently heat the oil bath while stirring the mixture. As soon as the two layers merge completely, stop heating and raise the boiling tube above the oil. Continue stirring briskly and note the temperature at which the solution becomes misty, *i.e.* at which two distinct liquid phases appear. Repeat with mixtures 5, 3, 6, 2, 7, 1 and 8 in this order. Mixtures 1, 2 and 8 may be found to be completely miscible at room temperature, but two layers appear during the initial stages of heating.

Insert the stirrer and the 0–110°C thermometer in mixture No. 4 and cool by immersing in iced water (in a 500-ml beaker). Stir continuously until complete miscibility is achieved. Remove the mixture from the cooling bath and continue stirring while the temperature rises, noting the temperature at which misting occurs. Repeat this procedure with mixtures 5, 3, 6 and 7. Mixtures already miscible at room temperature (probably 1, 2 and 8) require only to be warmed gently in the oil bath until misting occurs, the temperature being noted. Mixtures rich in glycerol are extremely viscous at the lower temperatures used and stirring must be done with care if very many air bubbles are not to be introduced. These bubbles interfere with observation of the misting temperature and their formation will be minimised if the surface is not broken with the stirrer ring.

Tabulate results, calculate the volume percentage of *m*-toluidine in the mixtures, plot a temperature–composition curve, and label the areas delineated by the curve. Note the upper and lower critical solution temperatures.

RESULTS

Tube No.	1	2	3	4	5	6	7	8
m-Toluidine (ml) Glycerol (ml)								
Vol. % *m*-toluidine								
Upper misc. temp. (°C) Lower misc. temp. (°C)								

Upper critical solution temperature = _____°C
Lower critical solution temperature = _____°C

THEORY

Read the theory concerning the phenol–water system—see Experiment 24.

Glycerol and *m*-toluidine are completely miscible at moderately high and moderately low temperatures, but form a two layer system over the intervening range. Thus the system has two critical solution temperatures—upper and lower. The temperature–composition curve is of the type shown below:

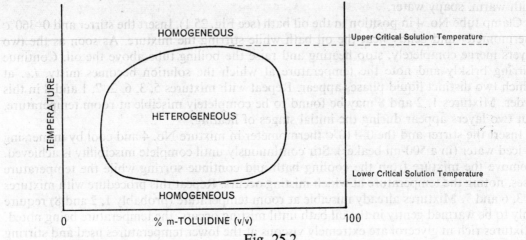

Fig. 25.2

EXPERIMENT 26. INVESTIGATION OF A TERNARY SYSTEM (ETHANOL–BENZENE–WATER)

APPARATUS AND CHEMICALS REQUIRED

Two 50-ml burettes; microburette (5 or 10 ml); three burette stands; 5-ml pipette; six stoppered bottles (preferably ca. 100 ml); triangular co-ordinate graph paper ('Chartwell' graph sheet no. 5590).

Ethanol (industrial methylated spirit) (90 ml); benzene (60 ml).

EXPERIMENTAL PROCEDURE

Charge one clean, dry 50-ml burette with ethanol and another with benzene. Charge the microburette with distilled water.

N.B. Benzene is carcinogenic, and every care must be taken to avoid inhaling the vapour or spilling the liquid on the skin.

Label the bottles 1 to 6, and prepare in them the following ethanol–benzene mixtures:

Bottle No.	1	2	3	4	5	6
Ethanol (ml)	2·5	7·6	12·7	17·7	22·8	24·1
Benzene (ml)	20·5	15·9	11·4	6·9	2·3	1·2

Titrate each solution very carefully with distilled water, shaking vigorously between additions, until the onset of turbidity. Note the volumes of water required (these will be very small for the first two or three titrations, so care must be taken to avoid running in too much water). As a precaution against adding too much water, keep back 1 or 2 ml of the ethanol–benzene mixture in a small pipette, to be added in the later stages of the titration.

Tabulate results and calculate the volume percentages of the components present at the onset of turbidity. Plot the percentage compositions on triangular co-ordinate graph paper (see p. 114) and label the areas delineated by the curve.

Record the room temperature and enter this on the graph.

RESULTS

Room temperature = _____°C

Mixture	ETHANOL		BENZENE		WATER		Total volume (ml)
	Volume (ml)	Volume %	Volume (ml)	Volume %	Volume (ml)	Volume %	
1							
2							
3							
4							
5							
6							

THEORY

The liquid pairs ethanol–benzene and ethanol–water both exhibit complete miscibility, but benzene and water are only slightly miscible. This experiment follows the varying degrees of miscibility when all three are present.

A given mixture of ethanol–benzene can mix with only a definite quantity of water (at a given temperature) without becoming turbid. Such ternary mixtures may be represented on triangular co-ordinate graph paper to give a curve of the type shown (Fig. 26.1).

Other systems which are easily investigated are aniline–*n*-butanol–water and benzene–acetic acid–water.

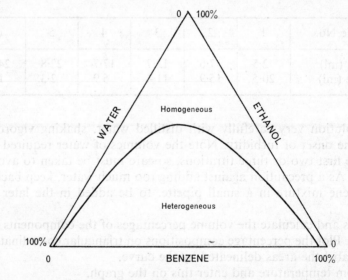

Fig. 26.1

EXPERIMENT 27. TO PREPARE AND DETERMINE THE COMPOSITION OF A CONSTANT BOILING MIXTURE OF HYDROGEN CHLORIDE AND WATER

APPARATUS AND CHEMICALS REQUIRED

500-ml distillation flask (preferably borosilicate glass); Liebig condenser and tubing; adaptor; 0–110°c thermometer (in 1°c); two retort stands and clamps; bunsen; 250-ml

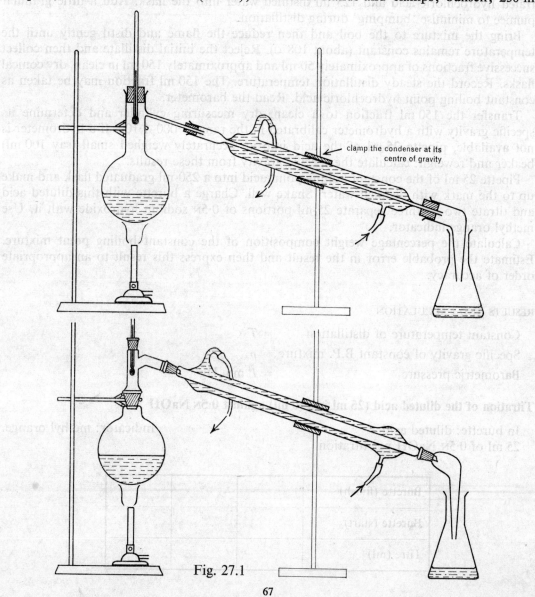

clamp the condenser at its centre of gravity

Fig. 27.1

measuring cylinder; burette and stand; white tile; 25-ml pipette; 250-ml graduated flask; four 250-ml conical flasks; wash bottle; hydrometer (S.G. range 1·000–1·100).

Concentrated hydrochloric acid (125 ml); 0·5M (0·5N) sodium hydroxide (100 ml); methyl orange indicator; pumice, granular (2 g).

EXPERIMENTAL PROCEDURE

Assemble the distillation apparatus (Fig. 27.1) (the bottom of the flask should be about 5 cm above the top of the bunsen, and should be heated direct) and pour 125 ml concentrated hydrochloric acid and 125 ml distilled water into the flask. Add a little granular pumice to minimise 'bumping' during distillation.

Bring the mixture to the boil and then reduce the flame and distil gently until the temperature remains constant (about 108°C). Reject the initial distillate and then collect successive fractions of approximately 50 ml and approximately 150 ml in clean, dry conical flasks. Record the steady distillation temperature. The 150 ml fraction may be taken as constant boiling point hydrochloric acid. Read the barometer.

Transfer the 150 ml fraction to a clean, dry measuring cylinder and determine its specific gravity with a hydrometer calibrated in the range 1·000–1·100. If a hydrometer is not available, pipette 25 ml of the acid into an accurately weighed small (say 100 ml) beaker and reweigh; calculate the specific gravity from these results.

Pipette 25 ml of the constant boiling point acid into a 250-ml graduated flask and make up to the mark with distilled water. Shake well. Charge a burette with this diluted acid and titrate two or three separate 25-ml portions of 0·5N sodium hydroxide with it. Use methyl orange indicator.

Calculate the percentage weight composition of the constant boiling point mixture. Estimate the probable error in the result and then express this result to an appropriate order of accuracy.

RESULTS AND CALCULATION

Constant temperature of distillation = $T°$C

Specific gravity of constant B.P. mixture = ρ

Barometric pressure   = P mm Hg

Titration of the diluted acid (25 ml in 250 ml) against 0·5N NaOH

In burette: diluted acid.   Indicator: methyl orange.
25 ml of 0·5N NaOH per titration.

Burette (finish)			
Burette (start)			
Titre (ml)			

_____ ml of diluted acid ≡ 25 ml 0·5N sodium hydroxide

∴ Normality of diluted acid $= $ _____ $N = N'$

∴ Normality const. B.P. acid $= 10 \times N' = N_m$

Now 1·0N hydrochloric acid contains 36·47 g HCl per 1000 ml solution.

∴ N_m hydrochloric acid contains $36·47 \times N_m$ g HCl per 1000 ml solution.

∴ % HCl (w/w) in constant boiling point mixture is equal to

$$\frac{36·47 \times N_m}{1000 \times \rho} \times 100\% = \text{_____} \% \text{ at } P \text{ mm Hg}$$

THEORY

In the hydrogen chloride–water mixture both components exhibit negative deviations from Raoult's law and the vapour pressure–composition curve passes through a minimum point as in Fig. 27.2 and the temperature–composition vaporus and liquidus curves are of the type shown in Fig. 27.3. A maximum boiling point mixture of composition Y is formed, the vaporus and liquidus curves coinciding at Y in the same manner as at the composition extremes, *i.e.* where there are single, pure components.

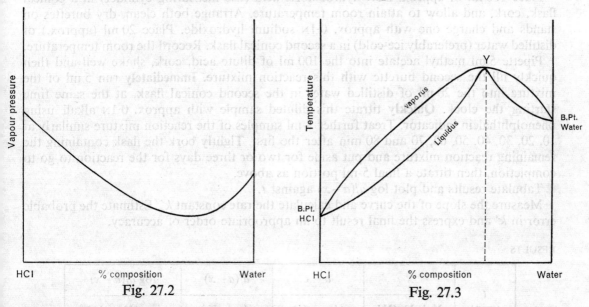

Fig. 27.2   Fig. 27.3

If a mixture richer in hydrogen chloride than Y is distilled the vapour is richer in hydrogen chloride than the liquid from which it came. Hence the composition of the liquid in the flask approaches Y and on reaching this point distils unchanged at a constant temperature (provided pressure is a constant), since the liquid and the vapour in equilibrium with it have the same composition. Similarly, if a mixture richer in water than Y is distilled the vapour is richer in water than the liquid; the liquid composition moves towards Y and on reaching it a constant boiling point liquid will distil.

The latter process is preferred for the preparation of Y, since no hydrogen chloride escapes into the laboratory.

Chemical Kinetics

EXPERIMENT 28. THE ACID CATALYSED HYDROLYSIS OF METHYL ACETATE, TO DETERMINE THE RATE CONSTANT

APPARATUS AND CHEMICALS REQUIRED

Two 50-ml burettes; two burette stands; white tile; 5-ml pipette; two 250-ml conical flasks (one with cork); 250-ml graduated flask; 250-ml beaker; 100-ml and 25-ml measuring cylinders; funnel; clock.

Methyl acetate (5 ml); approx. 0·5M (0·5N) hydrochloric acid (11 ml conc. acid made up to 250 ml); approx. 0·1M (0·1N) sodium hydroxide (500 ml); phenolphthalein indicator.

EXPERIMENTAL PROCEDURE

Place 100 ml of approx. 0·5N hydrochloric acid (use measuring cylinder) in a conical flask, cork, and allow to attain room temperature. Arrange both clean, dry burettes on stands and charge one with approx. 0·1N sodium hydroxide. Place 20 ml (approx.) of distilled water (preferably ice-cold) in a second conical flask. Record the room temperature.

Pipette 5 ml methyl acetate into the 100 ml of dilute acid, cork, shake well and then quickly fill the second burette with this reaction mixture. **Immediately** run 5 ml of the mixture into the 20 ml of distilled water in the second conical flask, at the same time starting the clock. **Quickly** titrate this diluted sample with approx. 0·1N alkali using phenolphthalein indicator. Treat further 5-ml samples of the reaction mixture similarly at 10, 20, 30, 40, 50, 60, 70 and 80 min after the first. Tightly cork the flask containing the remaining reaction mixture and put aside for two or three days for the reaction to go to completion, then titrate a final 5-ml portion as above.

Tabulate results and plot $\log a/(a-x)$ against t.

Measure the slope of the curve and calculate the rate constant k'. Estimate the probable error in k' and express the final result to an appropriate order of accuracy.

RESULTS

t (min)	x Vol. NaOH (ml)(V_t)	$a-x$ $(V_\infty - V_t)$ (ml)	$a/(a-x)$ $\dfrac{(V_\infty - V_0)}{(V_\infty - V_t)}$	$\log a/(a-x)$ $\log \dfrac{(V_\infty - V_0)}{(V_\infty - V_t)}$
0	(V_0)			
10				
20				
30				
⋮				
∞	(V_∞)			

$$a \equiv (V_\infty - V_0) = \underline{\hspace{2cm}} \text{ml}$$
$$\text{Room temperature} = \underline{\hspace{1.5cm}}°\text{C}$$

Concentration of methyl acetate = 5 ml per 105 ml solution $= \underline{\hspace{1.5cm}}$M

Concentration of catalyst (hydrogen ions from hydrochloric acid) = approx. 0·5M

Concentration of water, in large excess, is effectively constant

Rate constant $k' = 2·303 \times$ slope $= 2·303 \times \underline{\hspace{2cm}}$ $= \underline{\hspace{1.5cm}}$ min^{-1}

THEORY

$$CH_3COOCH_3 + H_2O \xrightarrow{H^+} CH_3COOH + CH_3OH$$

Let a and b represent the initial molarities of methyl acetate and water respectively, and x be the number of moles per litre of methyl acetate (and therefore of water) reacted after a time t.

Then the rate of the reaction is given by

$$\frac{dx}{dt} = k(a-x)(b-x) \quad \text{(where } k = \text{rate constant)}$$

Integration gives

$$t = \frac{1}{k(a-b)} \ln \frac{b(a-x)}{a(b-x)}$$

and this is the equation for a second order reaction. If, however, $b \gg a$ (as in this experiment) the equation may be rewritten

$$t = \frac{1}{-kb} \ln \frac{a-x}{a}$$

or, as b is also effectively constant,

$$t = \frac{2·303}{k'} \log \frac{a}{a-x}$$

where k' is a new rate constant.

This equation is identical with the kinetic equation for a first order reaction.

Plotting $\log a/(a-x)$ against t should give a straight line of slope $k'/2·303$.

Now V_0 is the volume of alkali required to neutralise the hydrochloric acid present—which remains constant;

V_∞ is the volume of alkali required to neutralise the hydrochloric acid and the acetic acid from the complete hydrolysis of the methyl acetate;

and V_t is the volume of alkali required to neutralise the hydrochloric acid and the acetic acid from the partial hydrolysis of the methyl acetate.

So

$$x \equiv V_\infty - V_0$$
$$a-x \equiv V_\infty - V_t$$

The reaction is said to be first order since the rate of reaction at any time t is dependent only upon the concentration of one reactant, the methyl acetate.

The samples taken during the experiment are run into a larger volume of cold water in order to 'freeze' the reaction long enough for the titrimetric analysis.

EXPERIMENT 29. THE HOMOGENEOUS CATALYTIC DECOMPOSITION OF HYDROGEN PEROXIDE

(To demonstrate that this is a first order reaction, and to investigate the effect of catalyst concentration.)

APPARATUS AND CHEMICALS REQUIRED

Magnetic stirrer; 250-ml Buchner flask as a reaction vessel; gas burette (a tapless 50-ml burette is suitable); levelling reservoir; rubber and glass tubing as shown; small trough

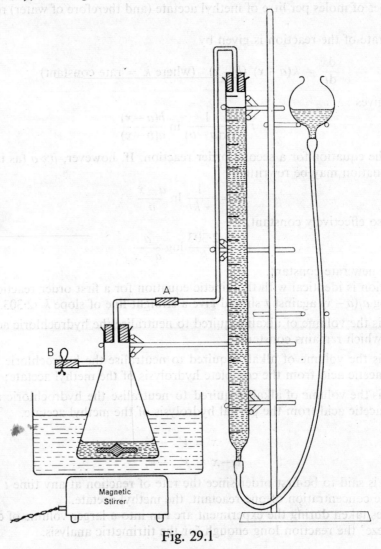

Fig. 29.1

filled with water and allowed to attain room temperature; retort stand, clamps and rings; 25-ml pipette; 100-ml graduated flask; 250-ml beaker; clock with second hand; spring clip.

20 volume hydrogen peroxide (4 ml); *freshly* prepared approx. M/3 ferric chloride (9 g hydrate made up to 100 ml of solution).

(A mechanical stirrer may be used in this experiment but this makes the apparatus more cumbersome.)

EXPERIMENTAL PROCEDURE

The apparatus is assembled as in Fig. 29.1 with the bottom of the reaction vessel about 2 cm above the bottom of the trough.

Disconnect the rubber joint A and remove the bung from the reaction vessel. Pipette 25 ml of M/3 ferric chloride and 2 ml of 20 volume hydrogen peroxide into the flask and replace the bung. Reconnect A, open clip B and adjust the water level in the burette to zero.

Close B, switch on the stirrer, and immediately start the clock. Note the volume of gas evolved after 2, 4, 6, 10, 15, 20, 25, 30, 45 and 60 min, and also when the reaction is complete (about $1\frac{1}{2}$ hours), equating levels on each occasion. It is advisable to move the reservoir down in step with the water level in the burette, since keeping the pressure within the apparatus at about that of the atmosphere minimises inaccuracies due to leakage.

The stirrer must rotate fairly briskly and at constant rate throughout the experiment.

Repeat the experiment using M/6 ferric chloride (pipette 50 ml of the M/3 solution into a 100-ml graduated flask and make up to the mark with distilled water). It is unnecessary to proceed to V_∞ in the second determination as this is the same as for the first.

Tabulate results and plot $\log (V_\infty - V_t)$ against t.

Compare the slopes of the curves obtained with the concentrations of catalyst used.

RESULTS

Reactant: 2 ml 20 volume hydrogen peroxide.   Catalyst: 25 ml of M/3 ferric chloride.

Time (t min)	Volume oxygen (V_t) ml	$(V_\infty - V_t)$	$\log (V_\infty - V_t)$
0			
2			
4			
6			
etc.			
	V_∞		

Record your results for the second reaction (using M/6 ferric chloride) similarly.

$$\frac{\text{Slope of graph using M/3 ferric chloride}}{\text{Slope of graph using M/6 ferric chloride}} = \frac{\quad}{\quad} = \frac{\quad}{1}$$

THEORY

The kinetic equation for a first order reaction is

$$t = \frac{1}{k} \ln \frac{a}{a-x}$$

which, in this case, may be written

$$t = \text{const.} \log \frac{1}{V_\infty - V_t}$$

Thus a plot of $\log (V_\infty - V_t)$ against t should give a straight line of negative slope if the reaction is kinetically of the first order.

The slopes of the straight lines should be proportional to the concentrations of catalyst used, in this case 2:1.

NOTE

If a suitable stirrer is not available the experiment may be performed using a manganese dioxide catalyst in place of the ferric chloride. The results are only slightly less satisfactory; the major difficulty is that the heterogeneous catalyst becomes insufficiently agitated during the later stages of the reaction. Use 1·0 g of catalyst for the first run through, and 0·5 g for the second.

EXPERIMENT 30. TO INVESTIGATE THE RELATIONSHIP BETWEEN REACTION VELOCITY AND TEMPERATURE, AND TO DETERMINE THE ENERGY OF ACTIVATION OF THE PERSULPHATE–IODIDE REACTION

$$S_2O_8^{2-} + 2I^- \rightarrow 2SO_4^{2-} + I_2$$

APPARATUS AND CHEMICALS REQUIRED

Four 250-ml conical flasks; two 25-ml pipettes; 5-ml pipette; pneumatic trough; two retort stands and clamps; 0–110°C in 1°C thermometer; stop clock; pipette safety filler.

0·05M (0·1N) potassium persulphate (200 ml) (or the sodium salt); 10% potassium iodide solution (200 ml); approximately 0·1M (0·1N) sodium thiosulphate solution (50 ml); starch indicator; ice.

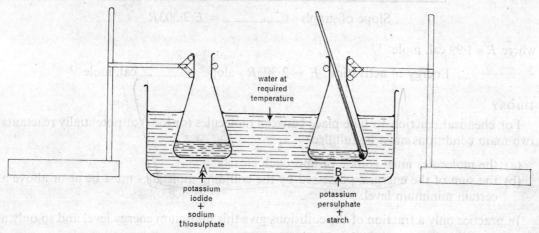

water at
required
temperature

A

B

potassium
iodide
+
sodium
thiosulphate

potassium
persulphate
+
starch

Fig. 30.1

EXPERIMENTAL PROCEDURE

Two-thirds fill a pneumatic trough with water at about 40°C (± 2°C). Clamp two clean, well-drained, conical flasks into position in the trough as shown. Pipette 25 ml 10% potassium iodide and 5 ml of approx. 0·1N sodium thiosulphate into flask A and 25 ml 0·1N potassium persulphate and a few drops of starch indicator into flask B. Place the thermometer in B. Leave the flasks for about 5 min to attain thermal equilibrium.

Unclamp A and quickly pour its contents into B, at the same time starting the clock. Agitate the reaction mixture well and leave immersed in the water bath. Note the time (t sec) at which the mixture suddenly turns blue-black and record its temperature.

Repeat the experiment several times using a cooler (eventually ice-cold) water bath on each occasion. Try to reduce the bath temperature by about 5°C each time by the addition of cold water.

Plot log t against $1/T$, where T is the absolute temperature. Use the graph to calculate the activation energy for the reaction. Estimate the probable error in the result and then express this result to an appropriate order of accuracy.

RESULTS

Reaction mixture on all occasions:

25 ml 0·1N potassium persulphate

25 ml 10% potassium iodide

5 ml 0·1N sodium thiosulphate

Time (t sec)	Temp. (°C)	Temp. (°K)	$\frac{1}{T}$	log t

Slope of graph = _____ = $E/2{\cdot}303R$

where $R = 1{\cdot}99$ cal. mole^{-1}

∴ Energy of activation, $E = 2{\cdot}303R \times$ slope = _____ cal. mole^{-1}

THEORY

For chemical reaction to take place between molecules (or ions) of potentially reactants two main conditions must be fulfilled:

(a) the molecules must collide or approach closely;
(b) the sum of the energies possessed by the colliding molecules must be at or above a certain minimum level.

In practice only a fraction of the collisions give this minimum energy level and so only a fraction of the collisions result in reaction.

Increasing the temperature of the reactants increases the reaction rate since

(a) the velocities of the molecules, and therefore the collision rate, increase;
(b) the average energy possessed by each molecule increases and so a greater fraction of the collisions have the required minimum energy level.

This minimum energy is called the 'energy of activation' for the reaction.

The relationship between temperature and reaction rate is given by the Arrhenius equation

$$k = A\,e^{-E/RT}$$

where $k =$ velocity constant, $T =$ absolute temperature, $E =$ energy of activation, $R =$ gas constant and $A = a$ constant.

Taking logs $\ln k = B - E/RT$ $(B = \text{constant})$

Examination of the kinetic equations for reactions of different orders indicates that if a particular reaction is observed at different temperatures (the initial conditions as regards reactant concentrations being standardised) then the velocity constant k is inversely proportional to the time t taken for a given fraction of the reaction to be completed, *i.e.*

$$k = \frac{\text{const.}}{t}$$

Hence $\qquad\qquad \ln k = C - \ln t \qquad (C = \text{constant})$

and therefore $\qquad\qquad \ln t = \dfrac{E}{RT} - D \qquad (D = \text{constant})$

or $\qquad\qquad \log t = \dfrac{E}{2 \cdot 303 RT} - D$

Thus a plot of $\log t$ against $1/T$ should give a straight line of slope $E/2 \cdot 303 R$, and hence E may be calculated.

In this experiment the reaction

$$S_2O_8{}^{2-} + 2I^- \rightarrow 2SO_4{}^{2-} + I_2$$

is conducted in the presence of thiosulphate ions, and the reaction

$$2S_2O_3{}^{2-} + I_2 \rightarrow S_4O_6{}^{2-} + 2I^-$$

removes iodine as it is formed until all the thiosulphate has reacted. At this point the iodine turns the starch blue.

Provided the volume of thiosulphate solution used is always the same for each run-through of the reaction, the blue coloration appears when the same fraction of the reaction is complete in each case. The presence of the thiosulphate renders the reaction first order, regardless of the relative concentrations of the persulphate and iodide, since the iodide ions are continuously regenerated. This is equivalent to conducting the experiment with a large excess of iodide ions.

Surface Phenomena

EXPERIMENT 31. INVESTIGATION OF THE ADSORPTION OF OXALIC ACID ON CHARCOAL

APPARATUS AND CHEMICALS REQUIRED

Six stoppered bottles (150- or 250-ml); two burettes and stands; white tile; 5-ml, 10-ml and 25-ml pipettes; three 250-ml conical flasks; two filter funnels; Whatman No. 1 filter papers (or equivalent) (12·5 cm); labels; thermostat if available.

Activated charcoal (granular) (30 g); approx. 0·25M (0·5N) oxalic acid (500 ml); 0·1M (0·1N) sodium hydroxide (250 ml); phenolphthalein indicator.

EXPERIMENTAL PROCEDURE

Weigh out six 5-g portions of activated charcoal (to within 0·01 g) and place them in dry stoppered bottles. Label the bottles A to F.

Charge one burette with distilled water and the other with approx. 0·5N oxalic acid. Run into each bottle the volumes of water and acid specified below.

Bottle	A	B	C	D	E	F
Water (ml)	0	20	40	60	80	90
Acid (ml)	100	80	60	40	20	10

Shake each mixture well and leave overnight, preferably in a thermostat at 20 or 25°C, so that equilibrium is established.

Wash the burettes and recharge one with 0·1N sodium hydroxide. Titrate a 5-ml portion of the original (approx. 0·5N) oxalic acid with this alkali, using phenolphthalein indicator.

Calculate the molarity of the original oxalic acid solution and the molarities of the acid solutions prepared in each bottle. Tabulate the results (Table 2).

When equilibrium has been established between the charcoal and the acid solutions, filter the contents of bottle A through the Whatman No. 1 paper, rejecting the first 5 ml approximately (because of adsorption of solute by the paper) and collecting the remainder in a *dry* conical flask. Pipette a 10-ml portion of this filtrate into another conical flask and titrate with the 0·1N alkali using phenolphthalein indicator. Repeat with the contents of the other five bottles, taking 10-ml portions of the filtrates from B and C and 25-ml portions from D, E and F. Tabulate results (Table 1) and calculate the molarity of oxalic acid in each of the filtrates.

Calculate the number of moles of acid adsorbed in each case (x) and plot graphs of

(i) x against concentration (c) of the solution in equilibrium with the charcoal,

(ii) $\log x$ against $\log c$, and from the slope calculate n.

Estimate the error in n and then express the result to an appropriate order of accuracy.

78

RESULTS

Standardisation of original (approx. 0·5N) oxalic acid

5 ml oxalic acid soln. ≡ _____ ml 0·1N sodium hydroxide

Normality of acid = _____N   Molarity of acid = _____M

Titration of filtrates from acid/charcoal equilibria

In burette: 0·1N sodium hydroxide.   Indicator: phenolphthalein.

TABLE 1

Bottle	A	B	C	D	E	F
Vol. filtrate used (ml) (*a*)	10	10	10	25	25	25
Burette (finish)						
Burette (start)						
Titre (ml) (*b*)						

The molarity of each filtrate is calculated:

$$a \text{ ml oxalic acid filtrate} \equiv b \text{ ml } 0·1\text{N alkali}$$

$$\text{Molarity} = \frac{b}{a} \times \frac{0·1}{2} \text{ M} \quad \text{(oxalic acid being dibasic)}$$

TABLE 2

Bottle	A	B	C	D	E	F
Water (ml)	0	20	40	60	80	90
Acid (ml)	100	80	60	40	20	10
Initial molarity						
Final (filtrate) molarity (*c*)						
Decrease in molarity (*d*)						
Moles adsorbed (*d*/10) (*x*)						

Weight of activated charcoal in each bottle   = 5·00 g

Thermostat (or room) temperature (as applicable) = _____°C

THEORY

If a gas or liquid is in contact with a solid, then the concentration of fluid molecules at the interface is higher than in the bulk. The fluid molecules are said to be adsorbed on to the surface of the solid. For granular adsorbents of high surface area (*e.g.* activated charcoal) the degree of adsorption at a given temperature may be represented by the empirical equation

$$\frac{x}{m} = kc^n$$

commonly known as the Freundlich isotherm, where x is the mass of adsorbed substance, m the mass of adsorbent, c the equilibrium concentration (in solution) of the adsorbed substance, and k and n are constants. This equation is also found to apply to the solute component of a solution in contact with a solid.

In this experiment the mass of adsorbent m is kept constant, and thus a plot of x against c should give a curve of the type shown in Fig. 31.1.

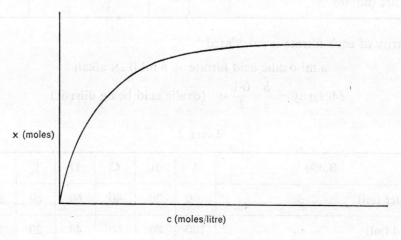

Fig. 31.1

The above graph shows how the amount adsorbed increases with concentration for low concentrations but tends to a limiting value for higher concentrations.

Taking logarithms of the equation we get

$$\log x - \log m = \log k + n \log c$$

and as m, k and n are constant a plot of $\log x$ against $\log c$ should give a straight line of slope n as in Fig. 31.2.

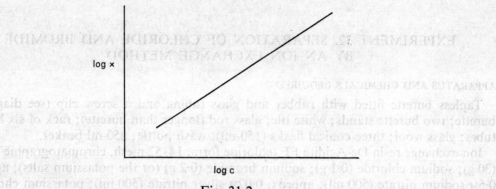

log x

log c

Fig. 31.2

EXPERIMENTAL PROCEDURE

Convert the resin in the column to the nitrate form by washing through with approx. 0·5M sodium nitrate until a sample of the effluent solution gives no precipitate with silver nitrate (about 125 ml of sodium nitrate solution is required). The resin will bed down considerably during this washing. Drain the solution to within 0·5 cm of the resin surface.

(3) Separation of chloride and bromide

Weigh about 0·1 g sodium chloride and about 0·2 g sodium bromide into a test-tube and dissolve in about 2 ml distilled water. Transfer this solution to the top of the

EXPERIMENT 32. SEPARATION OF CHLORIDE AND BROMIDE
BY AN ION EXCHANGE METHOD

APPARATUS AND CHEMICALS REQUIRED

Tapless burette fitted with rubber and glass tubing and a screw clip (see diagram); burette; two burette stands; white tile; glass rod (longer than burette); rack of six boiling tubes; glass wool; three conical flasks (150-ml); wash bottle; 250-ml beaker.

Ion exchange resin De-Acidite FF (chloride form, 14–52 mesh, chromatographic grade) (30 g); sodium chloride (0·1 g); sodium bromide (0·2 g) (or the potassium salts); approx. 0·5M sodium nitrate (500 ml); approx. 0·02M silver nitrate (500 ml); potassium chromate indicator.

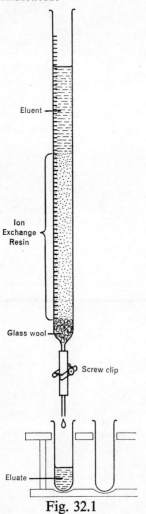

Eluent

Ion Exchange Resin

Glass wool

Screw clip

Eluate

Fig. 32.1

EXPERIMENTAL PROCEDURE

(1) Packing the column

Insert a substantial glass wool plug into the tapless burette and push it to the bottom with a long glass rod. Clamp the burette over the sink and fill to the brim with distilled water. Weigh about 30 g of the ion exchange resin into a 250-ml beaker and add about 100 ml distilled water. Stir and allow the resin to swell (5 min). Decant off most of the water and add the resin portionwise to the water-filled burette, allowing the water to overflow during the entire addition. (Packing the column in this way gives an even bed, free from air bubbles.) Run water from the column until it is about 0·5 cm from the resin surface. During the experiment the water or solution must never be allowed to fall below the resin surface, since the inclusion of air interferes with the functioning of the column.

(2) Preparing the column for use

Convert the resin in the column to the 'nitrate form' by washing through with approx. 0·5M sodium nitrate until a sample of the effluent solution gives no precipitate with silver nitrate (about 125 ml of sodium nitrate solution is required). The resin will bed down considerably during this washing. Drain the solution to within 0·5 cm of the resin surface.

(3) Separation of chloride and bromide

Weigh about 0·1 g sodium chloride and about 0·2 g sodium bromide into a test-tube and dissolve in about 2 ml distilled water. Transfer this solution to the top of the column and drain to within 0·5 cm of the surface. Now carefully top up the burette with 0·5M sodium nitrate and elute with this at a rate of about 2 ml per minute. Collect the eluate in 15-ml samples in boiling tubes (the volume collected being read from the graduations in the upper half of the burette). Wash the samples into conical flasks and carefully titrate each with approx. 0·02M silver nitrate, using potassium chromate indicator. Continue until a total of 180 ml eluate has been collected and titrated. (The elution may be arrested for up to a week without unduly affecting the results. Cork the burette during a prolonged arrest.) Tabulate results and plot the silver nitrate titres against total eluate volume.

RESULTS

15 ml samples of eluate titrated with 0·02N silver nitrate.

Total eluate (ml)	Burette (finish)	Burette (start)	Silver nitrate titre (ml)
15			
30			
45			
⋮			
180			

THEORY

An ion exchange resin consists essentially of large organic skeletal structures (polymers) which are electrically charged and bonded electrovalently to small ions of opposite charge. The resin used in this experiment may be represented by the structure

where the carbon skeleton contains both aliphatic and aromatic residues. This is an anion exchange resin. Cation exchange resins have negatively charged structures with positive ions bonded electrovalently. If an anion exchange resin is said to be in the 'chloride form' or 'nitrate form' this indicates the nature of the predominant anion. Similarly for cation exchange resins which may be encountered in the 'hydrogen' or 'sodium form', etc.

When an electrolyte solution is run through a bed of ion exchange resin (anion type, say) the anions in the solution interchange with those on the resin. For example, if sodium nitrate solution is run through an anion exchange resin in the 'chloride form' the effluent solution will contain sodium chloride. If the nitrate solution is very dilute then almost all nitrate ions will be exchanged for chloride ions during passage through the resin bed.

In this experiment the addition of a little chloride/bromide solution to the top of the 'nitrate form' resin converts a small band of the resin to a mixed 'chloride/bromide form' and elution with a nitrate solution causes this band to move down the column by a continuous ionic interchange process. The band not only broadens during its travel down the column but also becomes predominantly chloride at the leading end and bromide at the trailing end. Titration of eluate samples, and plotting an elution curve, indicates a considerable separation of the anions.

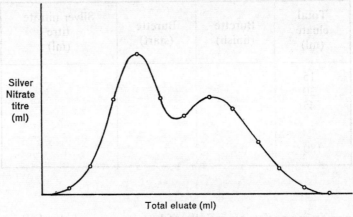

Fig. 32.2

EXPERIMENT 33. THE PREPARATION OF AN ARSENIOUS SULPHIDE SOL, AND ITS PRECIPITATION BY ELECTROLYTES

APPARATUS AND CHEMICALS REQUIRED

One 500-ml beaker; two 250-ml beakers; four 250-ml conical flasks; three burettes, with stands; black tile or similar dark background to place beneath titration flasks; 10-ml pipette; pipette safety filler; two 1000-ml, one 500-ml and one 250-ml graduated flasks; stirring rod; filter funnel (large); tripod; bunsen; wash bottle.

Kipp's apparatus for hydrogen sulphide *fitted with a wash bottle containing water*; arsenious oxide (1 g) (preferably amorphous, since it dissolves more quickly than crystalline); hydrated sulphates of sodium (7 g), magnesium (0·75 g), and aluminium (1·60 g).

EXPERIMENTAL PROCEDURE

Prepare the arsenious sulphide sol and solutions of the metal sulphates as follows:

(1) Arsenious sulphide sol

Add about a gram of arsenious oxide to about 300 ml of boiling distilled water and continue boiling for 15 min. Filter. Put aside about 100 ml of the solution and slowly pass water-washed hydrogen sulphide into the remainder until the solution just smells of the gas (take the solution well away from the gas supply to make this test **and take great care not to inhale the gas**). To the golden yellow sol obtained add more of the arsenious oxide solution (a little at a time, with stirring) until the smell of hydrogen sulphide disappears. Filter. A stable, golden-yellow sol of arsenious sulphide is obtained. Ignore a slight mistiness.

(2) Sodium sulphate ($Na_2SO_4,10H_2O$)

Dissolve 7·00 g (to 0·01 g) in water and make up to 250 ml. Label and note the concentration in grams per litre.

(3) Magnesium sulphate ($MgSO_4,7H_2O$)

Dissolve 0·75 g (to 0·01 g) in water and make up to 1000 ml. Label and note the concentration in grams per litre.

(4) Aluminium sulphate ($Al_2(SO_4)_3,18H_2O$)
(check bottle label for degree of hydration of sample used)

Dissolve 1·60 g (to 0·01 g) of the hydrate in water and make up to 1000 ml. Shake well and then pipette 10 ml of this solution into a 500-ml graduated flask and make up to the mark. Label this final solution, noting the concentration in grams per litre. Reject the first solution to avoid its use in error.

Titrate two or three separate 10-ml portions of the sol with each of the prepared electrolyte solutions until flocculation takes place. **Use a safety pipette filler for sampling**

7

the arsenious sulphide sol. A black surface beneath the titration flask will greatly assist the observation of the flocculation point. Perform the titrations slowly, shaking the flask well after each addition of electrolyte. Tabulate the results and calculate the mean titre for each electrolyte.

Calculate the metal ion concentration (in gram-ions per litre) required to precipitate the sol in each case, not forgetting to take account of the diluting effect of the 10-ml original volume of the sol. From these concentrations calculate the ratio of the ionic concentrations of the metal ions having the same precipitating effect, *i.e.* $[Al^{3+}]:[Mg^{2+}]:[Na^+]$, taking $[Al^{3+}]$ as unity.

RESULTS

Titration of arsenious sulphide sol with sodium sulphate solution

10 ml of sol used per titration. Self-indicating by flocculation.

Burette (finish)			
Burette (start)			
Titre (ml)			

Titration of arsenious sulphide sol with magnesium sulphate solution

Tabulate results as above.

Titration of arsenious sulphide sol with aluminium sulphate solution

Tabulate results as above.

Conc. of Na^+ required to precipitate sol   =   g.ions/litre

Conc. of Mg^{2+} required to precipitate sol =   g.ions/litre

Conc. of Al^{3+} required to precipitate sol =   g.ions/litre

Ratio $[Al^{3+}]:[Mg^{2+}]:[Na^+]$ = _____ :   :

THEORY

A partial explanation of the widely differing concentrations of 1-, 2- and 3-valent ions required to precipitate the same sol may be given in terms of the classical Freundlich isotherm (see Experiment 31).

Colloidal particles are electrically charged and a sol is precipitated when its particles adsorb oppositely charged ions from solution in sufficient quantity to neutralise their own electric charges enough to allow collisions to occur. From this one would expect the precipitating ions to be effective in direct proportion to the charges they carry, *i.e.* $Al^{3+}:Mg^{2+}:Na^+ = 3 : 1.5 : 1$, or Al^{3+} three times as effective as Na^+ and Mg^{2+} twice as effective as Na^+ in precipitating a negatively charged sol.

Assuming all ionic species to be adsorbed to the same extent (this is only very approximately true) and also that the adsorption is governed by the classical isotherm it can be

appreciated why the ratio of the ionic concentrations $[Al^{3+}]:[Mg^{2+}]:[Na^+]$ required to give the same sol precipitating power is of the order $\quad 1 \quad : \quad 10\text{–}20 \quad : \quad 700\text{–}1200.$
    Consider the classical isotherm below

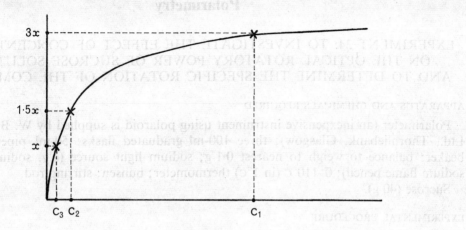

Fig. 33.1

The numbers of 3-, 2-, 1-valent ions adsorbed by the same quantity of sol to effect precipition are represented by x, $1\cdot5x$ and $3x$ respectively. From this it can be seen that the equilibrium concentrations of these ions c_3, c_2 and c_1 required to produce the adsorbed quantities increase rapidly with diminishing ionic charge.

Polarimetry

EXPERIMENT 34. TO INVESTIGATE THE EFFECT OF CONCENTRATION ON THE OPTICAL ROTATORY POWER OF SUCROSE SOLUTION, AND TO DETERMINE THE SPECIFIC ROTATION OF THE COMPOUND

APPARATUS AND CHEMICALS REQUIRED

Polarimeter (an inexpensive instrument using polaroid is supplied by W. B. Nicholson Ltd., Thornliebank, Glasgow); three 100-ml graduated flasks; 50-ml pipette; 250-ml beaker; balance to weigh to nearest 0·1 g; sodium light source (*e.g.* sodium lamp or sodium flame pencil); 0–110°C (in 1°C) thermometer; bunsen; stirring rod.

Sucrose (40 g).

EXPERIMENTAL PROCEDURE

(1) Determination of the 'zero' of the polarimeter

Using a 20-cm polarimeter tube, rinse this with distilled water and then fill with water until standing above the brim. Slide the circular optical glass cover slip into position over the end of the tube such that air bubbles are excluded, and carefully screw on the brass end-cap. Dry the outside of the tube, particularly the 'windows' in the end-caps, and place in position in the polarimeter. With the sodium source in position rotate the analyser while looking through the eyepiece and observe the changes in light intensity in the two semicircular areas in the field of view. Now set the analyser in the position giving equal darkness to the two areas. Note the reading on the analyser scale. This is the 'zero' of the instrument.

(2) Examination of sucrose solutions

Weigh out 40 g of sucrose (to within 0·1 g). Dissolve this in water and make up to 100 ml of solution. Shake well and label with the concentration. Use part of this solution to prepare solutions containing 20 g of sucrose/100 ml and 10 g/100 ml by successive dilution using a 50-ml pipette and two other 100-ml flasks. Allow all three solutions to attain room temperature and then determine the optical rotation of each in turn in the following way. Rinse out the 20-cm polarimeter tube with a little of the solution and then fill, close, *wash* and dry as before. Place the tube in the polarimeter and rotate the analyser to the new position of equal darkness. Note the degrees rotation from the determined 'zero' and denote by a (+) sign if to the right and a (−) sign if to the left. (There will in fact be two positions of equal darkness in 180° opposition, but by convention the rotation less than 90° is the one adopted.) Tabulate results. Note the room temperature.

Repeat the above procedure but starting with 30 g of sucrose and diluting to give solutions containing 15 g/100 ml and 7·5 g/100 ml. At the end of the experiment remove the end-cap, wash and dry the cap and tube, and leave the cap *off*. The cap can easily become corroded on before the apparatus is used again.

Complete the results table and plot a graph of 'degrees rotation per decimetre of solution' against 'conc. in g/ml'. The slope of the curve is then measured (this should approximate to a straight line). This is equal to the specific rotation of sucrose of units 'degrees.decimetre^{-1}.grams^{-1}.millilitres.' Estimate the error in the slope of the curve obtained and then express the specific rotation to an appropriate order of accuracy.

RESULTS

Zero reading of polarimeter = _____°

Solution	1	2	3	4	5	6
Concentration (g/100 ml)	40	20	10	30	15	7·5
Concentration (g/ml)	0·4	0·2	0·1	0·3	0·15	0·075
Rotation per 20 cm						
Rotation per 10 cm						

Room temperature   = ____°C

Wavelength of light used   = sodium D

From graph: Specific rotation = ____° decimetre^{-1}.g^{-1}.ml

THEORY

Some compounds in solution have the power of rotating the plane of polarised light. Such compounds are said to be optically active and, in the case of organic compounds, are found to have one or more asymmetric carbon atoms within a molecule. (An asymmetric carbon atom is one with four different groups attached to it).

The degree of rotation of the plane of polarised light is dependent on:

(a) the nature of the compound,
(b) the nature of the solvent,
(c) the depth of the solution through which the light passes,
(d) the concentration of the solution,
(e) the temperature,
(f) the wavelength of the light used.

The wavelength of the light normally used is that of the sodium D line, and in expensive apparatus the temperature is able to be kept constant at 20 or 25°C by use of a water jacket.

The specific rotation (α) of a given compound is given by

$$[\alpha]_\lambda^t = \frac{\text{degrees rotation per decimetre of solution}}{\text{grams of substance per ml of solution}}$$

where t is the temperature at which measured and λ is the wavelength of light used.

Optical rotation is measured using a polarimeter. Such an instrument is represented diagrammatically below.

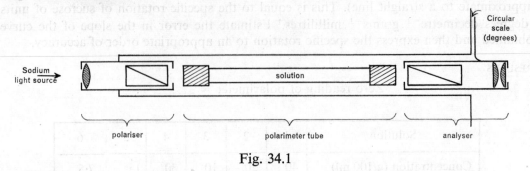

Fig. 34.1

There are three essential parts:

(i) the polariser (a lens and nicol prism assembly to polarise the incident light);
(ii) the polarimeter tube (a thick-walled glass tube with screw-on ends incorporating optical glass windows) which contains the solution under examination;
(iii) the analyser (a nicol prism and eyepiece assembly for 'crossing' the plane of the emergent polarised light); this is fitted with a degree scale and pointer.

The whole is enclosed in a light-excluding case. Inexpensive instruments are frequently fitted with 'polaroid' in place of nicol prisms. The prisms or polaroid are usually arranged so that the observed light is cut vertically into two distinct semicircular areas. One of these areas appears light and the other dark (Fig. 34.2, A). By rotating the analyser the intensities of the two areas may be made to approach and eventually become equal (Fig. 34.2, B). Further rotation in the same direction causes the light and dark areas to be reversed (Fig. 34.2, C) with a noticeable 'switch' from (A) through (B) to (C).

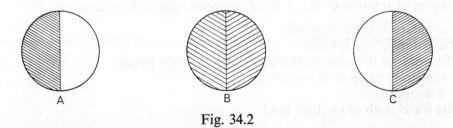

Fig. 34.2

The position of equal darkness (B) represents exclusion of polarised light, *i.e.* the polariser and analyser are 'crossed'. This position of equal darkness is adopted as the more reproducible one for taking readings.

Polarimetry may be used to determine the concentrations of solutions of optically active substances or the relative proportions of substances in optically active mixtures. It may also be used to study the kinetics of reactions involving optically active substances, (*e.g.* the kinetics of sucrose inversion as in the next experiment).

EXPERIMENT 35. TO VERIFY THE LAW OF MASS ACTION BY OBSERVING THE RATE OF THE ACID CATALYSED INVERSION OF SUCROSE

APPARATUS AND CHEMICALS REQUIRED

Polarimeter; 250-ml conical flask and cork; 100-ml measuring cylinder; sodium light source (sodium lamp or sodium flame pencil); 0–110°c (in 1°c) thermometer; clock; bunsen.
Sucrose (50 g); conc. hydrochloric acid (20 ml).

EXPERIMENTAL PROCEDURE

Weigh about 50 g sucrose (to 0·1 g), transfer to a conical flask and dissolve in about 75 ml water. If the flask is warmed to assist dissolving then the solution must be allowed to cool to room temperature before proceeding.

Using a 20-cm tube prepare the polarimeter to receive the reaction mixture (see Experiment 34 for method of filling the tube) and have the sodium light source ready. Record the room temperature.

Add about 20 ml of conc. hydrochloric acid to the sucrose solution, cork, shake well and immediately fill the polarimeter tube with this reaction mixture. Quickly wash and dry the outside of the tube (particularly the small 'windows' in the end-caps) and determine the optical rotation (θ_0), at the same time starting the clock. Leave the reaction mixture in the polarimeter and take further readings (θ_t) every 5 min until $t = 80$ min and then every 10 min until $t = 120$ min. Take a final reading (θ_∞) at $t = 150$ min when the reaction should be observationally complete.

Tabulate results and plot θ_t against t. Determine the slope of the curve ($d\theta_t/dt$) at several points (not less than six) by drawing tangents. (The section of the curve between $\theta_t = +10°$ and $\theta_t = -10°$ will be found best for this.) Then plot $d\theta_t/dt$ against θ_t, i.e. rate of change of concentration against concentration. Plot also $\log(\theta_t - \theta_\infty)$ against t.

RESULTS

M.W. sucrose = 342

$$\text{Approx. molar ratio } \frac{\text{sucrose}}{\text{water}} = \frac{50}{342}\Big/\frac{90^*}{18} \simeq \frac{1}{34}$$

Time (t min)	Optical rotation (θ_t deg.)	$\dfrac{d\theta_t}{dt}$ (from graph)	$(\theta_t - \theta_\infty)$	$\log(\theta_t - \theta_\infty)$
0 5 10 15 etc.		for a few values of θ_t and obtained from tangents to θ_t v t curve		

Room temperature = _____°C

* 75 ml (g) distilled water + approx. 15 g water in 20 ml conc. hydrochloric acid.

THEORY

Sucrose is dextrorotatory and as the reaction

$$C_{12}H_{22}O_{11} + H_2O \xrightarrow{\text{H}^+} C_6H_{12}O_6 + C_6H_{12}O_6$$
$$\text{(sucrose)} \qquad\qquad \text{(glucose)} \quad \text{(fructose)}$$

proceeds the optical rotation becomes less and eventually becomes negative (inverts). (Take care to use the appropriate $(+)$ and $(-)$ signs when noting rotations to the right and left respectively.) This is because (a) the sucrose concentration, and therefore its dextrorotatory power, decreases and (b) the laevorotatory power of fructose is greater than the dextrorotatory power of glucose. Thus θ_t may be taken as proportional to the sucrose concentration at time t. According to the law of mass action, the lower the concentration of sucrose the slower will be the rate at which it reacts. Thus a curve obtained by plotting θ_t against t should be of the type:

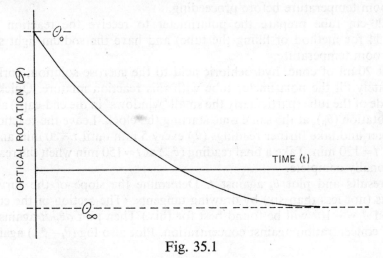

Fig. 35.1

If the slope $(d\theta_t/dt)$ of this curve is measured at several points, by drawing tangents, and $d\theta_t/dt$ (which represents the rate of reaction of sucrose) is plotted against θ_t, the curve obtained should be a straight line if the law of mass action holds good.

Another method of analysing the results may be adopted.

Although there are two reactants, one of them (water) is in appreciable molar excess, and therefore its concentration will remain approximately constant. Thus one would expect the reaction to be kinetically first order. The first order kinetic equation is

$$t = \frac{2\cdot303}{k} \log \frac{a}{a-x} \quad \text{(see Experiment 28)}$$

which in this case may be written

$$t = \frac{2\cdot303}{k} \log \frac{(\theta_0 - \theta_\infty)}{(\theta_t - \theta_\infty)}$$

A plot of $\log (\theta_t - \theta_\infty)$ against t should give a straight line of negative slope if the law of mass action (from which the kinetic equation is derived) holds good.

Electrochemistry

EXPERIMENT 36. TO DETERMINE (a) THE CONSTANT OF A CONDUCTANCE CELL AND (b) THE SPECIFIC AND EQUIVALENT CONDUCTANCES OF 0·1M (0·1N) HYDROCHLORIC ACID

The experimenter should first read the Appendix dealing with the types of circuit used to determine electrolytic conductance (p. 115).

APPARATUS AND CHEMICALS REQUIRED

Conductance circuit, including a dip type conductance cell; 100-ml borosilicate beaker or a boiling tube; 500-ml graduated flask; 250-ml beaker; filter funnel; stirring rod; wash bottle; thermostat (if available); 0–110°C (in 1°C) thermometer.

Potassium chloride (preferably 'AnalaR' grade) (4 g); 0·1M (0·1N) hydrochloric acid; conductance water (500 ml) (or specially distilled water may be used if conductance water is not available—see Appendix (iii)).

EXPERIMENTAL PROCEDURE

Prepare a 0·1M solution of potassium chloride (3·730 g per 500 ml solution) using conductance water (if available).

Connect the dip cell into the chosen conductance circuit and wash by dipping into about 50–100 ml of the 0·1M potassium chloride solution and agitating. Then dip the cell into a fresh portion (approx. 50 ml) of the potassium chloride solution in a dry 100-ml borosilicate glass beaker or boiling tube, the latter standing in a thermostat (if available). Allow to attain thermal equilibrium and then obtain the balance point in the manner appropriate to the circuit being used. With earphone detectors the balance point is at minimum noise level, and with an oscilloscope it is indicated by the disappearance of the sine waveform to give a straight line trace. Record the temperature of the solution.

Wash the cell thoroughly with distilled water and then repeat the above procedure using 0·1M (0·1N) hydrochloric acid in place of the potassium chloride solution.

Record the resistance of the cell filled with each solution and calculate first the cell constant and then use this to calculate the specific and equivalent conductances of the 0·1N hydrochloric acid. Estimate the probable error in the final results and then express them to an appropriate order of accuracy.

RESULTS AND CALCULATION

The specific conductance of 0·1M potassium chloride solution as determined by Kohlrausch at different temperatures is

$$10°C \quad = 0·00934 \text{ ohm}^{-1}.\text{cm}^{-1}$$
$$18°C \quad = 0·01120 \text{ ohm}^{-1}.\text{cm}^{-1}$$
$$25°C \quad = 0·01289 \text{ ohm}^{-1}.\text{cm}^{-1}$$

Resistance of cell filled with 0·1M KCl = ____ ohm (R)

Temperature of this solution = ____ °C

Specific conductance of 0·1M KCl (κ_{KCl}) = const . × $1/R$ ohm^{-1}.cm^{-1}

Cell constant $(a) = R \times \kappa_{KCl}$

= ____

Resistance of cell filled with 0·1M HCl = ____ ohm (R_1)

Temperature of this solution = ____ °C

Specific conductance of 0·1M HCl (κ_{HCl}) = a/R_1 ohm^{-1}.cm^{-1}

= ____ ohm^{-1}.cm^{-1} (at ____ °C)

Equivalent conductance of 0·1M HCl (Λ_{HCl}) = specific conductance × vol. in ml containing 1 g equiv.

= ____ ohm^{-1}.cm^2 equiv.$^{-1}$ (at ____ °C)

THEORY

The specific resistance of a medium is the resistance per cm length per cm^2 cross-section of the medium. Specific conductance is the reciprocal of this and is expressed as ohm^{-1} (reciprocal ohms, mhos). The specific conductance of a solution in a conductivity cell could be calculated from the measured cell resistance and a knowledge of the cell's precise dimensions. This would be a tedious process of doubtful accuracy. Instead, the cell is first calibrated using a solution of accurately known specific conductance. Then

$$\kappa = ax = a/R \quad \text{ohm}^{-1}.\text{cm}^{-1}$$

where κ = specific conductance, x = measured conductance (*i.e.* R = measured resistance) and a = a constant (the 'cell constant'). Once the cell constant is known the cell may be used for the accurate determination of the specific conductance of any other solution simply by multiplying the reciprocal of the resistance of the solution in this cell by the cell constant.

The equivalent conductance of a solution (Λ) is defined as the product of the specific resistance of the solution and the volume in ml containing 1 gram-equivalent of the electrolyte, *i.e.*

$$\Lambda = \kappa V \quad \text{ohm}^{-1}.\text{cm}^2 \text{ equiv.}^{-1}$$

EXPERIMENT 37. CONDUCTIMETRIC TITRATIONS

The experimenter should first read the notes on conductance circuits on p. 115.

APPARATUS AND CHEMICALS REQUIRED

Conductance circuit, including a dip type conductance cell; microburette (5 or 10 ml) and stand; 25-ml pipette; two 100-ml beakers; six borosilicate boiling tubes and rack; 10-ml measuring cylinder; piece of thin glass tubing (longer than the microburette).

0·1M (0·1N) hydrochloric acid (75 ml); 0·1M (0·1N) acetic acid (75 ml); 0·1M (0·1N) sodium chloride (25 ml); approx. 1·0M (1·0N) sodium hydroxide (25 ml); approx. 1·0M (1·0N) ammonium hydroxide (25 ml); solid silver nitrate (1·2 g).

EXPERIMENTAL PROCEDURE

In this experiment it is unnecessary to know the cell constant. In all of the titrations the solution added from the microburette must always be more concentrated (5 to 10 times) than that in the reaction vessel in order to avoid large volume changes during the titration.

(1) Titration of a strong acid with a strong base (HCl—NaOH)

Charge the microburette with approx. 1·0N sodium hydroxide. (If trapped air renders the filling of the burette difficult, the temporary insertion of a long piece of glass tubing will probably overcome the difficulty.)

Pipette 25 ml 0·1N hydrochloric acid into a clean, dry boiling tube. Immerse the dip cell in this solution and measure the resistance by obtaining a balance point with the chosen circuit. (With earphones the balance is at minimum noise level and with an oscilloscope it is indicated by the disappearance of the sine waveform to give a straight line trace.) Lift the cell from the solution, taking care to lose no drops, and run in 0·5 ml of approx. 1·0N sodium hydroxide from the microburette, agitating well. Return the cell, dipping it in and out of the solution several times before leaving it immersed, and measure the new resistance. Add further 0·5-ml portions of alkali in this manner, measuring the resistance each time, until a total of 5 ml has been added. Tabulate results. Thoroughly wash the cell with distilled water.

Plot a titration curve of $1/R$ against volume of alkali added, and use the graphical equivalence point to calculate the normality of the alkali. Estimate the probable error in the final result and then express the normality to an appropriate order of accuracy.

(2) Titration of a weak acid with a strong base (CH₃COOH—NaOH)

Repeat the above titration using 0·1N acetic acid in place of the hydrochloric acid.
Tabulate results and plot titration curves as above. Calculate the normality of the alkali.

(3) Titration of a strong acid with a weak base (HCl—NH₄OH)

Repeat the titration with approx. 1·0N ammonia in the microburette and 0·1N hydrochloric acid in the titration vessel.

Tabulate results, analyse them graphically, and calculate the normality of the ammonium hydroxide.

(4) Titration of a weak acid with a weak base (CH₃COOH—NH₄OH)

Repeat the titration with approx. 1·0N ammonia in the microburette and 0·1N acetic acid in the titration vessel.

Tabulate results, analyse them graphically, and calculate the normality of the ammonium hydroxide.

(5) Precipitation titration of a chloride solution with silver nitrate

Prepare an approx. 1·0N solution of silver nitrate by weighing out about 1·2 g (to 0·01 g) of the solid and dissolving in about 7 ml of distilled water in a test-tube. Charge a clean microburette with this solution after first washing out with about 1·5 ml of it.

Repeat the titration procedure using 25 ml of 0·1N sodium chloride in the titration vessel, and adding 0·5 ml portions of silver nitrate solution. Silver chloride is precipitated during the titration and at its completion the cell must be washed thoroughly with bench ammonia and then with distilled water to remove any of the precipitate which has adhered to the cell walls and electrodes.

Tabulate results, analyse them graphically, and calculate the normality of the prepared silver nitrate solution.

RESULTS

Record the results as indicated below for each titration.

Titration of . . . . with . . . .

In titration vessel: 25 ml . . . .

In burette: . . . .

Total volume of . . . . added (ml)	Resistance (R) of cell and contents (ohms)	$\frac{1}{R}$ = conductance (ohms^{-1})

From the graph:

25 ml . . . . ≡ . . . . ml of . . . .

Normality of . . . . =   = _____ N

THEORY

The conductance of an electrolyte solution (at constant temperature) depends upon (i) the ionic concentration (ii) the equivalent conductances of the different ions present. Most ions have equivalent conductances of the same order (see table on page 98) but that of the 'hydrogen' ion is approximately five times as great as most other ions and the hydroxyl ion three times as great.

Consider the titration of a strong acid with a strong base, *e.g.* HCl—NaOH.

On adding the alkali to the acid the 'hydrogen' ions (of high equivalent ionic conductance) are removed by the added hydroxyl ions to form the little dissociated water and are replaced by sodium ions of much lower equivalent ionic conductance:

$$H^+ + Cl^- + \underbrace{Na^+ + OH^-}_{added} \rightarrow Na^+ + Cl^- + H_2O$$

Thus the conductance falls until the equivalence point is reached (even though the ionic concentration remains constant).

After the equivalence point, the ionic concentration increases rapidly (since the added solution is more concentrated) and, moreover, the added hydroxyl ion has a high equivalent conductance. Thus the conductance increases sharply (Fig. 37.1).

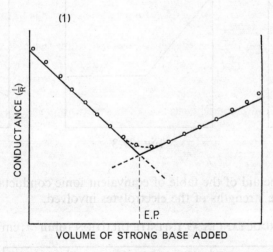

VOLUME OF STRONG BASE ADDED

Fig. 37.1

In the precipitation titration between silver nitrate and sodium chloride, on adding the silver nitrate, chloride ions are replaced in solution by nitrate ions

$$Na^+ + Cl^- + \underbrace{Ag^+ + NO_3^-}_{added} \rightarrow Ag^+Cl^- \downarrow + Na^+ + NO_3^-$$

and as these two ions have similar equivalent conductances the overall conductance changes little. After the equivalence point the added silver nitrate causes an increase in the ionic concentration and the conductance rises (Fig. 37.2).

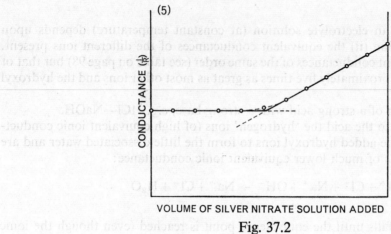

VOLUME OF SILVER NITRATE SOLUTION ADDED

Fig. 37.2

The curves of the other titrations are of the form:

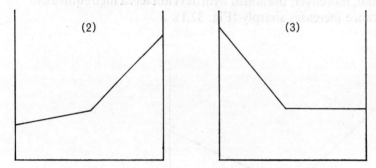
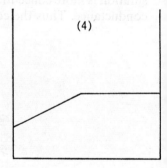

Fig. 37.3

Interpret these with the aid of the table of equivalent ionic conductances below and from your knowledge of the strengths of the electrolytes involved.

EQUIVALENT IONIC CONDUCTANCES AT INFINITE DILUTION (ohm^{-1}.cm^2) AT 18°C

Ion	E.I.C.	Ion	E.I.C.
H^+	315	OH^-	174
NH_4^+	64	Cl^-	65
Ag^+	54	NO_3^-	62
Na^+	43	CH_3COO^-	35

EXPERIMENT 38. POTENTIOMETRIC REDOX TITRATIONS

APPARATUS AND CHEMICALS REQUIRED

Slide wire potentiometer (1 metre); standard calomel electrode; electric stirrer; bright platinum electrode; sensitive galvanometer for use as a null instrument; 2v accumulator; voltmeter (0–3 v); potential divider (18 or 25 ohms); key; jockey; two 100-ml beakers; 250-ml beaker; 25-ml pipette; burette (with bent tip attachment as shown) and stand; salt bridge (see p. 121).

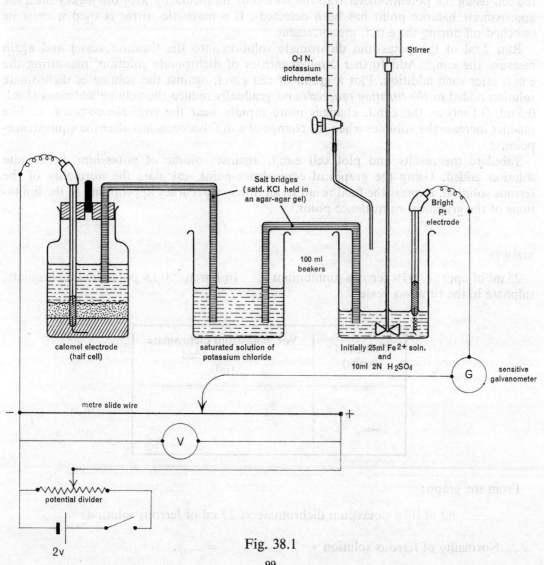

calomel electrode (half cell)

saturated solution of potassium chloride

Initially 25ml Fe²⁺ soln. and 10ml 2N H₂SO₄

Stirrer

O·I N. potassium dichromate

Salt bridges (satd. KCl held in an agar-agar gel)

Bright Pt electrode

100 ml beakers

G — sensitive galvanometer

metre slide wire

V

potential divider

2v

Fig. 38.1

0·1N (M/60) potassium dichromate (50 ml); approx. 0·1N (0·1M) ferrous ammonium sulphate (30 ml); dilute (bench) sulphuric acid (10 ml); saturated potassium chloride solution (50 ml).

EXPERIMENTAL PROCEDURE

Assemble the apparatus and prepare the circuit as shown in Fig. 38.1. Adjust the potential divider so that the voltage across the metre slide wire is 1 volt, then 1 mm ≡ 1 mv.

Pipette 25 ml of the approx. 0·1N ferrous solution into the clean titration vessel and add about 10 ml of bench dilute sulphuric acid. Switch on the stirrer. Measure the e.m.f. of the cell using the potentiometer (tap the wire only momentarily with the jockey until the approximate balance point has been detected). If a magnetic stirrer is used it must be switched off during the e.m.f. measurement.

Run 2 ml of the potassium dichromate solution into the titration vessel and again measure the e.m.f. Add further 2-ml quantities of dichromate solution, measuring the e.m.f. after each addition. Plot a graph of cell e.m.f. against the volume of dichromate solution added *as the titration proceeds* and gradually reduce the volume additions (1 ml, 0·5 ml, 0·1 ml) as the e.m.f. changes more rapidly near the equivalence-point. In like manner increase the volumes when the change of e.m.f. becomes less after the equivalence-point.

Tabulate the results and plot cell e.m.f. against volume of potassium dichromate solution added. Using the graphical equivalence-point, calculate the normality of the ferrous solution. Express the final result to an order of accuracy appropriate to the limitations of the graphical equivalence-point.

RESULTS

25 ml of approx. 0·1N ferrous ammonium sulphate in the titration vessel.   In burette: 0·1N potassium dichromate.

Cell e.m.f. (volts)	Vol. potassium dichromate solution added (ml)

From the graph:

_____ ml of 0·1N potassium dichromate ≡ 25 ml of ferrous solution

∴ Normality of ferrous solution =   = _____N

THEORY

The system

$$Fe^{3+} + e \rightleftharpoons Fe^{2+}$$

is a REDOX system, and its oxidising or reducing powers depend on the relative concentrations of the oxidant (Fe^{3+}) and reductant (Fe^{2+}) present.

If an inert electrode (*e.g.* bright platinum) is placed in a solution containing a redox system a p.d. is set up between the solution and the electrode, and is given by

$$E = E^0 + \frac{RT}{nF} \ln \frac{[\text{Oxidant}]}{[\text{Reductant}]} \text{ volts}$$

where E is the 'redox potential' for the system. $E = E^0$ when [Oxidant] = [Reductant] and E^0 is called the 'standard redox potential' for the system; n is the number of electrons involved in the equilibrium. In this experiment the equation becomes

$$E = E^0 + \frac{RT}{F} \ln \frac{[Fe^{3+}]}{[Fe^{2+}]} \text{ volts} \quad (n = 1)$$

E cannot be measured directly, but if a second, standard, electrode (*e.g.* a calomel electrode) is introduced to form a cell, then the cell e.m.f. may be measured potentio-

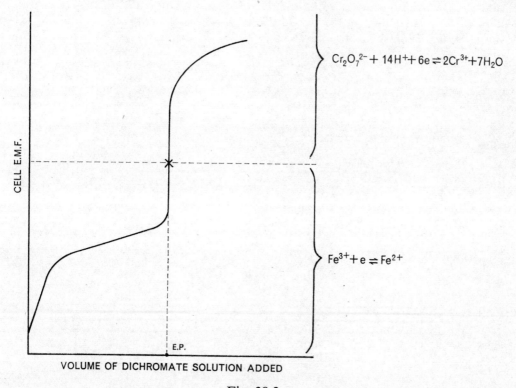

Fig. 38.2

metrically. The cell e.m.f. is thus a measure of the ratio $[Fe^{3+}]/[Fe^{2+}]$. This type of cell may be used to follow the course of a titration in which Fe^{2+} is being oxidised to Fe^{3+}.

During the titration of a ferrous ion solution with potassium dichromate there is an initial steep rise in the cell e.m.f. as the first small quantities of ferric ions are formed. There is another steep rise in e.m.f. as the equivalence-point is approached, *i.e.* as the last traces of ferrous ions are removed from the system. Beyond the equivalence-point the curve becomes the redox curve for the system

$$Cr_2O_7{}^{2-} + 14H^+ + 6e \rightleftharpoons 2Cr^{3+} + 7H_2O$$

(pH 2·68). Measure the e.m.f. of the cell. Make further additions of 0·2N sodium acetate (1 ml, 1·5, 2·5, 6·, 7 and 10 ml, using pipettes), measuring the e.m.f. after each
e.m.f. From the graph obtained write the equation for the cell showing the relationship between 7 and pH. Estimate the probable error in ...

Experimental Physical Chemistry

EXPERIMENT 39. TO INVESTIGATE THE RELATIONSHIP BETWEEN THE pH AND THE POTENTIAL OF A QUINHYDRONE ELECTRODE

APPARATUS AND CHEMICALS REQUIRED

Slide wire potentiometer (1 metre); wire jockey; sensitive centre reading galvanometer; key; potential divider (18 or 25 ohms); voltmeter (0–3 volts); single accumulator cell; standard calomel electrode; two 100-ml beakers; 2-ml graduated pipette; 10-ml graduated pipette; salt bridge; 0–110°C (in 1°C) thermometer; bright platinum electrode.

Quinhydrone (1 g); 0·2M (0·2N) acetic acid (15 ml); 0·2M (0·2N) sodium acetate (50 ml); saturated potassium chloride solution (50 ml).

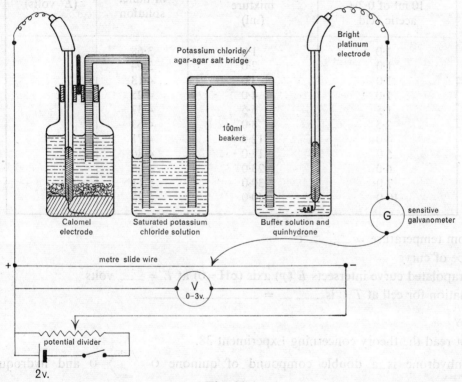

Fig. 39.1

EXPERIMENTAL PROCEDURE

Set up the cell and circuit shown. The standard calomel electrode is kept remote from the quinhydrone vessel to avoid contamination of the former. Adjust the potential divider until the p.d. across the metre slide wire is 0·5 v (*i.e.* 2 mm ≡ 1 mv).

Pipette 10 ml of 0·2M acetic acid into the 100-ml beaker, add a little quinhydrone (sufficient to just cover the finger nail) and swirl to dissolve. Now pipette 1 ml of 0·2M sodium acetate into the beaker and swirl to mix. This produces the first buffer solution (pH = 3·68). Measure the e.m.f. of the cell. Make further additions of 0·2M sodium acetate (1, 1, 1, 1·5, 2·5, 4, 5, 6, 7 and 10 ml, using pipettes), measuring the e.m.f. after each addition. Record the room temperature.

Plot a graph of cell e.m.f. against the pH of the buffer solution in the quinhydrone half-cell. From the graph obtained write the equation for the cell showing the relationship between E' and pH. Estimate the probable error in the slope of the curve and in the value of the intercept (A) obtained from it. Express the values of the slope and intercept to appropriate orders of accuracy.

RESULTS

Vol. 0·2M sodium acetate (ml) added to 10 ml of 0·2M acetic acid	Total vol. of 0·2M sodium acetate in mixture (ml)	pH of buffer solution	Cell e.m.f. (E' volts)
1·0	1·0	3·68	
1·0	2·0	3·94	
1·0	3·0	4·13	
1·0	4·0	4·25	
1·5	5·5	4·37	
2·5	8·0	4·53	
4·0	12·0	4·71	
5·0	17·0	4·86	
6·0	23·0	4·98	
7·0	30·0	5·11	
10·0	40·0	5·23	

Room temperature = _____°C

Slope of curve   =   =

Extrapolated curve intersects $E'(y)$ axis (pH = 0) at E' = _____ volts

Equation for cell at T°c is _____ = _____

THEORY

First read the theory concerning Experiment 38.

Quinhydrone is a double compound of quinone O=⟨ ⟩=O and hydroquinone HO—⟨ ⟩—OH and in aqueous solutions it dissociates to give equimolecular quantities of the single compounds:

$$C_6H_4O_2 . C_6H_4(OH)_2 \rightarrow C_6H_4O_2 + C_6H_4(OH)_2$$
quinhydrone   quinone   hydroquinone

Together with the hydrogen ions in the solution these compounds set up the redox system

$$\underbrace{C_6H_4O_2 + 2H^+ + 2e}_{\text{oxidant}} \rightleftharpoons \underbrace{C_6H_4(OH)_2}_{\text{reductant}}$$

and if the electrons are provided by a platinum electrode dipping into the solution the potential of the electrode (redox potential) E is given by

$$E = E^0 + \frac{RT}{2F} \ln \frac{[\text{quinone}][\text{H}^+]^2}{[\text{hydroquinone}]} \quad \text{volts}$$

where E^0, R and F are constants.

But since the molecular concentrations of quinone and hydroquinone are of necessity the same in this case, the equation simplifies to

$$E = E^0 + \frac{RT}{2F} \ln [\text{H}^+]^2 \quad \text{volts}$$

$$= E^0 + \frac{RT}{F} \ln [\text{H}^+] \quad \text{volts}$$

$$= E^0 + \frac{2\cdot303RT}{F} \log [\text{H}^+] \quad \text{volts}$$

giving

$$E = E^0 - 0\cdot058 \text{ pH} \quad \text{volts}$$

(at 20°c, having substituted for the constants R and F).

The p.d. between the platinum electrode and the quinhydrone solution is thus dependent on the pH of the solution. This arrangement is known as a 'quinhydrone electrode' or 'quinhydrone half-cell'. As in Experiment 38, the potential of the electrode cannot be measured direct. The measurement must be effected by the inclusion of another (standard) electrode to complete a cell, the e.m.f. of which may be measured potentiometrically. A calomel electrode may be conveniently used to complete the cell, and the cell e.m.f. (E') will be given by

$$E' = A - 0\cdot058 \text{ pH} \quad \text{volts} \quad . \quad . \quad . \quad . \quad . \quad \text{(i)}$$

where A is a constant which includes the potential of the calomel electrode.

In this experiment the mixtures of 0·2M acetic acid and 0·2M sodium acetate produce buffer solutions of known pH in the presence of quinhydrone. The experimenter should verify some of the pH values given, by calculation. (See theory of Experiment 11.)

A plot of cell e.m.f. (E') against the pH of the quinhydrone electrode should give a straight line (equation (i)) from which the constant A may be obtained. The slope of the curve will vary with temperature and so will not necessarily be 0·058.

The quinhydrone electrode behaves as a form of hydrogen electrode (for details of this consult a theoretical work) in that its potential varies with H^+ concentration. The hydrogen electrode is in many ways clumsy and difficult to use with success, mainly because a source of very pure hydrogen is essential and because it is easily 'poisoned'. The quinhydrone electrode has the following advantages:

(a) it is easy to set up and use;

(b) it attains a state of equilibrium very rapidly;

(c) it may be used in a wide variety of solutions without the risk of being 'poisoned'.

Its disadvantage is that it cannot be used to measure pH above 8–9 since

(i) it is a weak acid and begins to affect the pH itself, and
(ii) hydroquinone is readily oxidised in alkaline solutions and this disturbs the 1:1 ratio of   [quinone]:[hydroquinone].

In addition to being useful for measuring pH, the quinhydrone electrode may be used as a standard electrode (to replace, say, the calomel electrode) if buffered to a definite pH. A 0·05M solution of potassium hydrogen phthalate (pH = 4) is normally used for this latter purpose.

EXPERIMENT 40. POTENTIOMETRIC TITRATIONS USING A QUINHYDRONE ELECTRODE

APPARATUS AND CHEMICALS REQUIRED

Slide wire potentiometer (1 metre); wire jockey; sensitive galvanometer for use as a null instrument; potential divider (18 or 25 ohms); key; voltmeter (0–3 v) single accumulator; standard calomel electrode; two 100-ml beakers; 25-ml pipette; KCl salt bridge (see p. 121); bright platinum electrode; wire for circuit; burette and stand; electric stirrer.

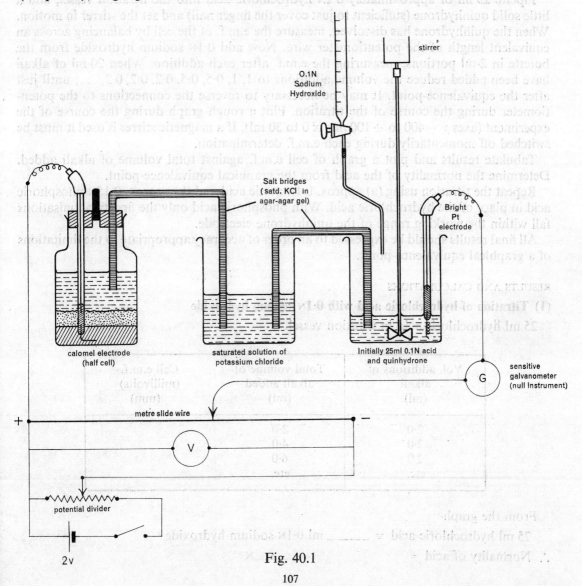

Fig. 40.1

0·1M (0·1N) sodium hydroxide (150 ml); 0·1M (0·1N) hydrochloric acid (30 ml); approx. 0·1M (0·1N) acetic acid (30 ml); approx. 0·05M phosphoric acid (1 ml of syrupy acid made up to about 300 ml with distilled water); saturated solution of potassium chloride (50 ml); quinhydrone (1 g).

EXPERIMENTAL PROCEDURE

Set up the cell and circuit shown in Fig. 40.1. The standard calomel electrode is kept remote from the titration vessel to avoid contamination of the former. Adjust the potential divider until the p.d. across the potentiometer wire is one volt (then 1 mm ≡ 1 mv).

Pipette 25 ml of approximately 0·1N hydrochloric acid into the titration vessel, add a little solid quinhydrone (sufficient to just cover the finger nail) and set the stirrer in motion. When the quinhydrone has dissolved, measure the e.m.f. of the cell by balancing across an equivalent length of the potentiometer wire. Now add 0·1N sodium hydroxide from the burette in 2-ml portions, measuring the e.m.f. after each addition. When 20 ml of alkali have been added reduce the volume additions to 1, 1, 0·5, 0·5, 0·2, 0·2, 0·2, . . ., until just after the equivalence-point. It may be necessary to reverse the connections to the potentiometer during the course of the titration. Plot a rough graph during the course of the experiment (axes y − 400 to + 100 mv: x 0 to 30 ml). If a magnetic stirrer is used it must be switched off momentarily during each e.m.f. determination.

Tabulate results and plot a graph of cell e.m.f. against total volume of alkali added. Determine the normality of the acid from the graphical equivalence-point.

Repeat the titration using (a) approx. 0·1N acetic acid and (b) approx. 0·15N phosphoric acid in place of the hydrochloric acid. With phosphoric acid only the first two ionisations fall within the working range of the quinhydrone electrode.

All final results should be expressed to an order of accuracy appropriate to the limitations of a graphical equivalence-point.

RESULTS AND CALCULATIONS

(1) **Titration of hydrochloric acid with 0·1N sodium hydroxide**

25 ml hydrochloric acid in titration vessel.

Vol. additions of alkali (ml)	Total volume of alkali added (ml)	Cell e.m.f. (millivolts) (mm)
2·0	2·0	
2·0	4·0	
2·0	6·0	
etc.	etc.	

From the graph

25 ml hydrochloric acid ≡ _____ ml 0·1N sodium hydroxide

∴ Normality of acid =   = _____N

(2) Titration of acetic acid with 0·1N sodium hydroxide

Enter results as above.

(3) Titration of phosphoric acid with 0·1N sodium hydroxide

Enter results as above.

(In this case the first inflexion represents one-third and the second inflexion two-thirds of the total equivalence.)

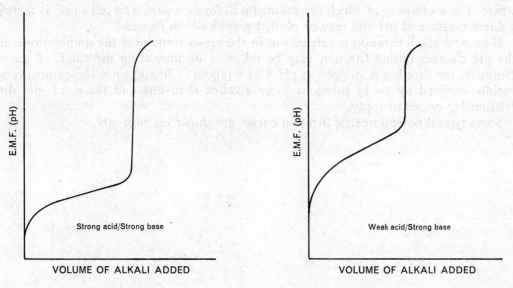

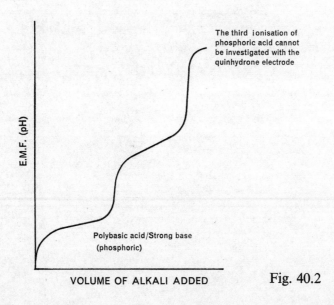

Fig. 40.2

THEORY

First read the theory applying to Experiment 38.

The potential of the calomel electrode is constant, but the potential of the quinhydrone electrode is dependent on the pH of the solution in which the quinhydrone is dissolved. Thus the e.m.f. of the cell is pH dependent and is given by the equation

$$E' = A - 0.058\ \text{pH}\quad\text{volts}$$

where A is a constant of which the calomel p.d. forms a part. The cell e.m.f. is therefore a direct measure of pH and may be plotted graphically in its stead.

If an acid-alkali titration is carried out in the vessel containing the quinhydrone, then the pH changes during titration may be followed by measuring the e.m.f. of the cell. Normally the titration is stopped at pH 8 or 9 (about $+70$ mv) since the quinhydrone is rapidly oxidised by air in solutions more alkaline than this and the e.m.f.–pH direct relationship ceases to apply.

Some typical potentiometric titration curves are shown on page 109.

Further Experiments

The experiments in this section are intended to allow the student to develop further the practical and deductive skills acquired during a course based on Experiments 1–40. Only a minimum of information is given, and before attempting any of the Experiments 41–50 the student should carefully plan the work and seek the approval of the tutor. The plan of the experiment might be of the form:

  (i) a list of apparatus, solutions and reagents required;
  (ii) a sketch of the apparatus (where applicable);
  (iii) brief notes on expected procedure;
  (iv) the measurements to be taken;
  (v) an indication of how the results might be analysed.

EXPERIMENT 41. (Ionic Equilibria and Solubility)

Investigate the effect of total ionic concentration on the 'solubility product' of calcium hydroxide. Use a graphical method to present your results.

Use solutions of sodium hydroxide varying in concentration from $1 \cdot 0$M ($1 \cdot 0$N) to $0 \cdot 01$M ($0 \cdot 01$N).

Relevant experiment in the main text: 3.

EXPERIMENT 42. (Phase Equilibria)

Investigate the aniline–n-hexane system in order to construct a phase diagram.
A water bath will provide a sufficiently high temperature.
Relevant experiments in the main text: 24 and 25.

EXPERIMENT 43. (Phase Equilibria)

Investigate the urea–phenol system in order to construct a phase diagram. Comment on the diagram.
A bath of liquid paraffin will be required for heating.
Relevant experiment in the main text: 23.

111

EXPERIMENT 44. (Phase Equilibria)

Investigate the effects (separately) of the presence of small quantities of potassium chloride, succinic acid and soap on the critical solution temperature for the phenol–water system. Account for your results.

As a guide, the concentrations of impurities should not exceed about 0.2 mole.litre^{-1} of water.

Relevant experiments in the main text: 24 and 25.

EXPERIMENT 45. (Thermochemistry: Heats of Ionisation)

Determine the heats of the three ionisations of phosphoric acid by a calorimetric method.

Bear in mind that the equilibrium

$$HPO_4^{2+} + OH^- \rightleftharpoons H_2O + PO_4^{3+}$$

lies well to the left, and so an excess of alkali must be used for the third determination. The heat of the reaction

$$H^+ + OH^- \rightarrow H_2O$$

may be taken as $\Delta H = -13,700$ calories.

Relevant experiments in the main text: 8, 9 and 10.

EXPERIMENT 46. (Thermochemistry: Heat of Solution)

Determine, in turn, the heats of solution of one mole of ethyl alcohol in 2.5, 6, 10, 20, 40 and 60 moles of water. Plot a graph of ΔH against 'number of moles of water' and comment on the curve obtained.

Use industrial methylated spirit for this experiment. Calculate the volumes of alcohol and water used for each determination so that the volume of the mixture in the calorimeter is about 100 ml.

Relevant experiments in the main text: 8, 9 and 10.

EXPERIMENT 47. (Chemical Kinetics: A Conductance Method)

Investigate the kinetics of the reaction

$$CH_3COOCH_2CH_3 + OH^- \rightarrow CH_3CH_2OH + CH_3COO^-$$

by conductance measurements.

Use equal volumes of 0·1M sodium hydroxide and 0·1M ethyl acetate for the reaction mixture. When the solutions are mixed, measure the conductance of the mixture at suitable periods within the total time of about one hour.

Plot conductance against time and interpret the curve obtained. Analyse the results in another way to elucidate the kinetics of the reaction.

Relevant experiments in the main text: 28, 29, 35, 36, 37 and Appendix (ii).

EXPERIMENT 48. (Conductance Measurements)

Determine, by titrating conductimetrically, the composition of the given mixture Y of concentrated hydrochloric acid and glacial acetic acid.

The concentrated mixture must first be appropriately diluted.

Relevant experiments in the main text: 36, 37 and Appendix (ii).

EXPERIMENT 49. (Potentiometric Measurement of pH)

Use a potentiometric method (quinhydrone electrode) to determine the dissociation constants of acetic, monochloroacetic, dichloroacetic and trichloroacetic acids, and account for your results.

Set up a cell using two quinhydrone electrodes and buffering one to pH 4 to provide a standard. E^0 for the quinhydrone electrode may be taken as $-0·711$ volts at 15°C.

Relevant experiments in the main text: 12, 39 and 40.

EXPERIMENT 50. (Partition Law)

Determine the partition coefficient for hydrogen chloride between benzene and water, using hydrochloric acid of different concentrations for each of several determinations. Comment on the results.

Use solutions of hydrochloric acid of the following approximate molarities: 10M, 5M, 2·5M, 1·0M. Take great care when equilibrating the 10M acid with benzene.

Relevant experiments in the main text: 4, 5, 6 and 7.

APPENDIX

(i) USE OF TRIANGULAR CO-ORDINATE GRAPH PAPER

Triangular graph paper may be used to represent the variations of concentration of the three components (*A*, *B* and *C*) of ternary systems at constant temperature and pressure.

Each corner represents a pure component, *e.g.* the apex *A* represents a system of pure *A*, *i.e.* 100% *A*.

Mixtures containing other percentages of *A* will be given on points on lines drawn parallel to the side opposite to apex *A* (Fig. 50.1).

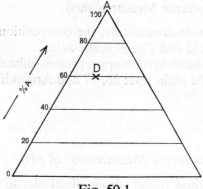

Fig. 50.1

Thus *D* represents a mixture containing 60% *A* (as well as some *B* and *C*).

The percentages of *B* and *C* are recorded in a similar fashion on lines parallel to the sides opposite to apexes *B* and *C* respectively.

Ternary mixtures will be represented by points lying within the triangle, thus a mixture consisting of 20% *A*, 60% *B* and 20% *C* will be given by the point *P* (Fig. 50.2).

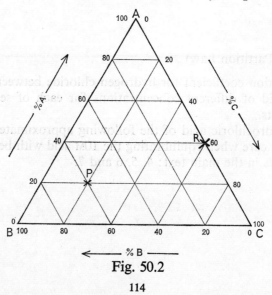

Fig. 50.2

114

A point on the side of the triangle represents a binary system containing none of the component represented by the opposite apex, *e.g.* R represents a mixture of 60% C and 40% A.

These graphs represent isothermal systems and if a third variable temperature is introduced the phase diagram becomes a solid figure, *viz.* a prism with an equilateral triangular base, the triangular graphs being sections taken parallel to the base at the appropriate temperatures (Fig. 50.3).

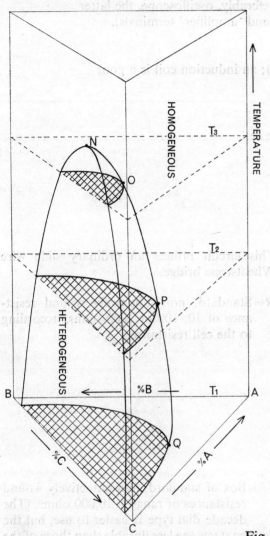

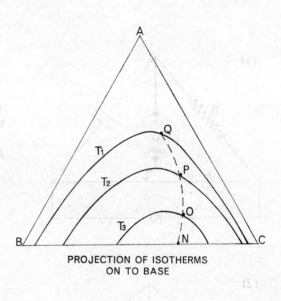

PROJECTION OF ISOTHERMS
ON TO BASE

Fig. 50.3

(ii) CIRCUITRY IN THE MEASUREMENT OF ELECTROLYTE CONDUCTANCE

Electrolyte conductance (reciprocal resistance) is almost always determined by including a cell, containing the electrolyte, in a Wheatstone bridge circuit and measuring its resistance.

Any of the following circuits may be used, but they are recommended in the order 4, 3, 2, 1, taking both accuracy and ease of operation into account.

In all the circuits:

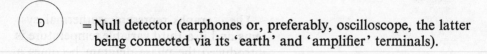

= Null detector (earphones or, preferably, oscilloscope, the latter being connected via its 'earth' and 'amplifier' terminals).

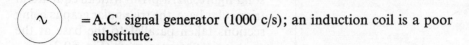

= A.C. signal generator (1000 c/s); an induction coil is a poor substitute.

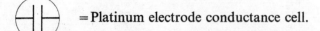

= Platinum electrode conductance cell.

(1)

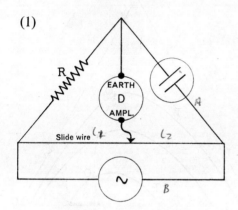

This circuit utilises an ordinary slide wire Wheatstone bridge.

R = Standard, non-inductively wound resistance of 10, 100 or 1000 ohms, according to the cell resistance.

(2)

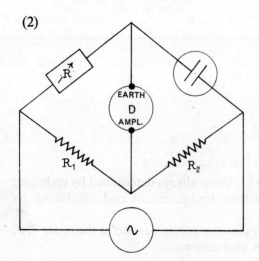

$R^{\nearrow}$ = Box of standard, non-inductively wound resistances of range 1–10,000 ohms. (The decade dial type is easier to use, but the contacts are less durable than those of the peg type box.)

$R_1 = R_2$ = Standard 100 or 1000 ohms non-inductively wound resistances.

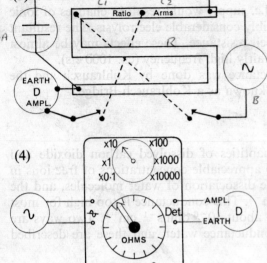

(3)

This circuit uses the familiar Post Office Box form of the Wheatstone bridge.

(4)

This 'circuit' shows one type of commercially produced conductance bridge provided with multiplier and a dial for the direct reading of resistance at the balance point.

Some instruments include a 'magic eye' detector, eliminating the necessity for earphones or oscilloscope, and also a fixed-frequency oscillator.

The conductance cell recommended for routine work in the student laboratory is the dipping type with platinum black electrodes. This is generally the most robust and most convenient to use. If a cell is new or has not been used for some time it should be immersed in distilled water for about 30 min before use.

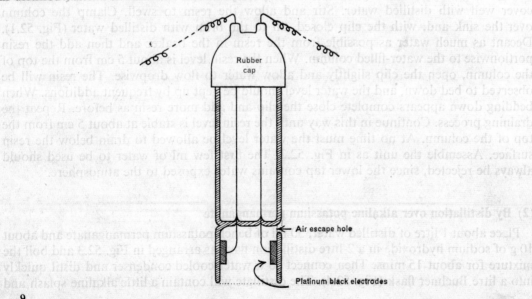

During the measurement of the resistance of an electrolyte-filled cell it is inevitable that it conducts a finite current, and the use of a d.c. supply would result in changes of ionic concentrations around the electrodes and possibly considerable electrolysis. The resultant polarisation would bring about a change of cell resistance. These effects may be almost entirely eliminated by using an a.c. supply of fairly high frequency (ca. 1000 c/s).

Much original work on electrolyte conductance was done by Kohlrausch and the above type of a.c. bridge circuit is commonly known as a Kohlrausch Bridge.

(iii) CONDUCTANCE WATER

Ordinary distilled water contains small quantities of dissolved carbon dioxide and ammonium salts. These impurities give rise to appreciable concentrations of free ions in comparison with those derived solely from the dissociation of water molecules, and the specific conductance is about 10^{-5} ohms^{-1}.cm^{-1}. This conductance is too high for most conductance work, and it must be reduced to about 10^{-6} ohms^{-1}.cm^{-1}. Two ways are ordinarily available for the preparation of conductance water, and these are described below.

(1) By using an ion exchange resin

The purchase of a commercial ion exchange unit is recommended if large quantities of conductance water are required, but a simple laboratory-constructed unit gives good results. Fig. 52.2 shows a suggested arrangement. All glass components must be of borosilicate glass (*e.g.* 'Pyrex') since ordinary soda glass is appreciably soluble and would defeat the whole object of the demineralising process.

The column should be packed with a mixed resin in the following way. Place the estimated quantity of resin ('Biodeminrolit' pH dye indicated form) in a litre beaker and cover well with distilled water. Stir and allow the resin to swell. Clamp the column over the sink and, with the clip closed, fill to the brim with distilled water (Fig. 52.1). Decant as much water as possible from the resin in the beaker and then add the resin portionwise to the water-filled column. When the resin level is about 5 cm from the top of the column, open the clip slightly and allow water to flow dropwise. The resin will be observed to bed down, and the water level should be kept up by frequent additions. When bedding down appears complete close the clip and add more resin as before. Repeat the draining process. Continue in this way until the resin level is stable at about 5 cm from the top of the column. At no time must the water level be allowed to drain below the resin surface. Assemble the unit as in Fig. 52.2. The first few ml of water to be used should always be rejected, since the lower tap contains water exposed to the atmosphere.

(2) By distillation over alkaline potassium permanganate

Place about 1 litre of distilled water, 50 ml of bench potassium permanganate and about 10 g of sodium hydroxide in a 2-litre distillation flask as arranged in Fig. 52.3 and boil the mixture for about 15 mins. Then connect to a water-cooled condenser and distil quickly into a litre Buchner flask (Fig. 52.4). The distillate will contain a little alkaline splash and

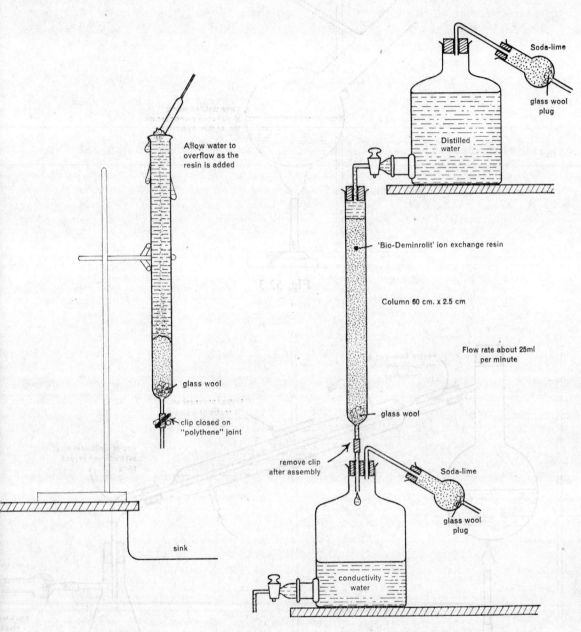

Allow water to
overflow as the
resin is added

glass wool

clip closed on
"polythene" joint

sink

Soda-lime

glass wool
plug

Distilled
water

'Bio-Deminrolit' ion exchange resin

Column 60 cm. x 2.5 cm

Flow rate about 25ml
per minute

glass wool

remove clip
after assembly

Soda-lime

glass wool
plug

conductivity
water

Fig. 52.1   Fig. 52.2

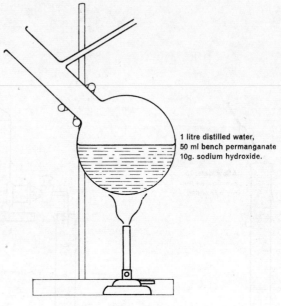

1 litre distilled water,
50 ml bench permanganate
10g. sodium hydroxide.

Fig. 52.3

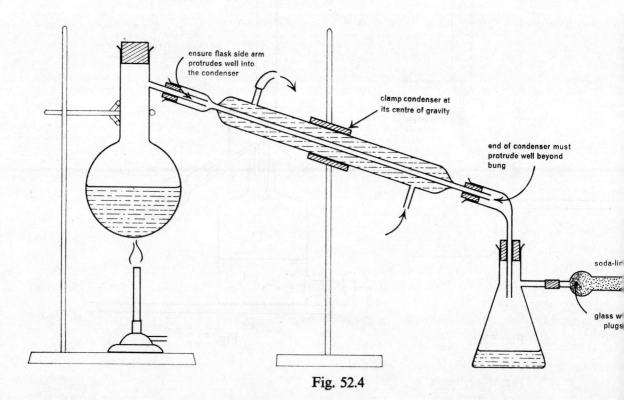

ensure flask side arm
protrudes well into
the condenser

clamp condenser at
its centre of gravity

end of condenser must
protrude well beyond
bung

soda-lir

glass w
plugs

Fig. 52.4

must be redistilled from a clean, dry, borosilicate flask and condenser. The water may be stored in borosilicate, stoppered bottles, filled to the brim to exclude air.

(iv) PREPARATION OF A SALT BRIDGE

APPARATUS AND CHEMICALS REQUIRED

100-ml beaker with lip; tripod; bunsen; stirring rod; glass tubing (bore ca. 5 mm); retort stands and clamps; batswing burner; rubber policemen.

Potassium chloride (16 g); agar-agar (1·5 g).

PROCEDURE

A salt bridge is used to establish an electrical connection between two solutions which for other reasons must be kept apart. A piece of wire could not be used because the potentials set up between the metal and the solutions would interfere with the e.m.f. measurements usually being made. A salt bridge often takes the form of an inverted U-tube which has been filled with a saturated solution of potassium chloride held in a 3% agar-agar gel.

(1) Dissolve 16 g of potassium chloride in about 50 ml of hot distilled water in a 100-ml (lipped) beaker. Add about 1·5 g agar-agar and boil gently until the latter has dissolved (at least 15 min) taking care that the solution does not froth over. Add a little water from time to time to make up for evaporation losses.

(2) While the solution is boiling, bend pieces of glass tubing to the required U-shapes and clamp them, limbs pointing upwards, with the ends at the same level (Fig. 53.1).

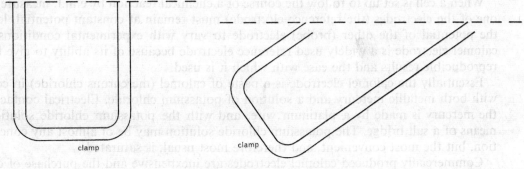

Fig. 53.1

(3) Fill the tubes with the hot potassium chloride–agar-agar solution by pouring directly from the beaker into one limb. As the solution cools in the tubes and sets to a gel it will shrink into the limbs slightly and a little more of the hot solution must be added to each end. When the tubes have cooled to room temperature the bridges are ready for use.

(4) After use the limbs of the bridges must be washed thoroughly with distilled water. They may be reserved for future use either by leaving the limbs immersed in saturated potassium chloride solution in test tubes or by closing the ends with rubber policemen (Fig. 53.2). The first method suffers from the disadvantage that a capillary rise effect results in a crystalline growth of potassium chloride 'creeping' out of the test-tubes. If

rubber policemen are used they should first be moistened and then pushed no more than about one-eighth of an inch onto the limbs; if pushed on further the gel will be forced out of the tube.

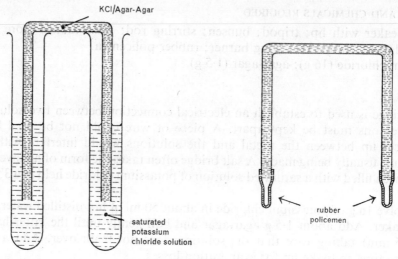

Fig. 53.2

(v) THE STANDARD CALOMEL ELECTRODE

When a cell is set up to follow the course of a chemical reaction by e.m.f. measurements one of the electrodes (the reference electrode) must remain at constant potential, leaving the potential of the other (probe) electrode to vary with experimental conditions. The calomel electrode is a widely used reference electrode because of its ability to give highly reproducible results and the ease with which it is used.

Essentially the calomel electrode is a paste of calomel (mercurous chloride) in contact with both metallic mercury and a solution of potassium chloride. Electrical contact with the mercury is made by a platinum wire, and with the potassium chloride solution by means of a salt bridge. The potassium chloride solution may be of almost any concentration, but the most convenient, and therefore most usual, is saturated.

Commercially produced calomel electrodes are inexpensive and the purchase of one or two is well worth while. The experimenter or laboratory technician may wish, however, to prepare one from materials readily available in the laboratory. The following procedure is recommended.

CHEMICALS, MATERIALS AND APPARATUS REQUIRED

Calomel (pure); potassium chloride; mercury (clean and free from dissolved metals); salt bridge prepared as on p. 121.

Small wide-necked bottle (50–60 ml) with triple-bored bung or cork to fit; glass tubing (bore approx. 5 mm); copper wire flex; 3 cm of platinum wire; glass rod; rubber policeman.

Pestle and mortar; filter funnel; three 100-ml beakers; 250-ml beaker; 100-ml measuring cylinder.

PROCEDURE

Clean and dry the bottle and pour in mercury to a depth of about 1 cm. Prepare a saturated solution of potassium chloride (warm 35 g with 100 ml water) and pour some of this onto the mercury until the bottle is half to two-thirds full. Make a thick paste of calomel in the mortar by mixing with a little of the saturated potassium chloride solution and grind in a little mercury. Place the calomel paste in the bottle and allow it to settle into

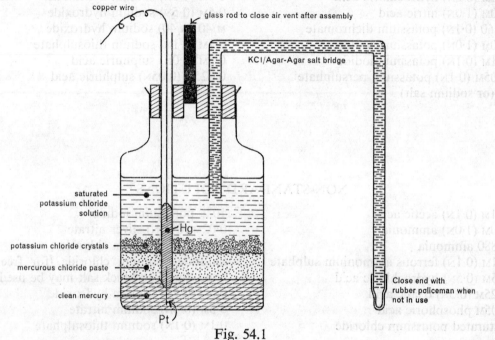

Fig. 54.1

a layer on top of the mercury, and to about an equal depth. When settled add 2–3 g of potassium chloride crystals.

Prepare a salt bridge as on p. 121.

Seal a 3-cm length of platinum wire into the end of a 12-cm length of glass tubing so that 0·5–0·75 cm protrudes outwards.

Moisten the holes of the bung and push the platinum probe and the salt bridge into position. The platinum probe must dip completely into the mercury and the salt bridge well into the potassium chloride solution when the bung is inserted in the bottle. With the bung in position in the bottle, insert a short length (2–3 cm) of glass rod in the third hole to isolate the inside of the bottle from the atmosphere and so avoid evaporation of the solution. Do not insert the rod more than a few mm or the pressure build-up will force the gel from the salt bridge. If the electrode is not to be used immediately, the open end of the salt bridge must be closed with a rubber policeman. When the cell is in use electrical contact with the mercury layer is made by pouring a little mercury into the platinum probe tube, followed by the insertion of a copper wire flex.

The saturated standard calomel electrode has a potential (hydrogen scale) of −0·245 volts at 20°C.

(vi) CHEMICALS AND STANDARD SOLUTIONS REQUIRED

STANDARD SOLUTIONS

0·2M (0·2N) acetic acid

0·1M (0·1N) hydrochloric acid ✓

0·2M (0·2N) hydrochloric acid ✓

1·0M (1·0N) nitric acid ✓

M/60 (0·1N) potassium dichromate

1·0M (1·0N) potassium iodide

0·1M (0·1N) potassium iodide ✓

0·05M (0·1N) potassium persulphate ✓
　(or sodium salt)

0·2M (0·2N) sodium acetate

1·0M (1·0N) sodium bicarbonate

1·0M (1·0N) sodium hydroxide ✓

0·5M (0·5N) sodium hydroxide

M/40 (N/40) sodium hydroxide ✓

0·1M (0·1N) sodium thiosulphate ✓

0·5M (1·0N) sulphuric acid ✓

0·025M (0·05N) sulphuric acid ✓

NON-STANDARD SOLUTIONS

0·1M (0·1N) acetic acid

1·0M (1·0N) ammonia

0·880 ammonia

0·1M (0·1N) ferrous ammonium sulphate

0·5M (0·5N) hydrochloric acid

0·25M (0·5N) oxalic acid

0·05M phosphoric acid

Saturated potassium chloride

10% potassium iodide ✓

0·02M (0·02N) silver nitrate

0·1M (0·1N) sodium chloride

Saturated sodium chloride (for freezing
　mixtures, so block salt may be used)

0·5M (0·5N) sodium hydroxide

0·5M (0·5N) sodium nitrate

0·1M (0·1N) sodium thiosulphate

INORGANIC REAGENTS

aluminium sulphate (hydrate)

ammonium nitrate

arsenious oxide (pref. amorphous)

calcium hydroxide

ferric chloride (hydrate)

hydrochloric acid (conc.)

hydrogen peroxide (20 vol.)

iodine

magnesium sulphate (hydrate)

mercuric iodide

mercurous chloride

phosphoric acid

potassium chloride ('AnalaR' grade)

potassium nitrate

potassium permanganate

silver nitrate

sodium bicarbonate

sodium bromide

sodium carbonate (decahydrate)

sodium chloride

sodium sulphate (decahydrate)

ORGANIC REAGENTS

acetic acid (glacial)
acetone
benzene
carbon tetrachloride
chloroform
chlorobenzene
citric acid (pure cryst.)
dichloroacetic acid
ethyl acetate
ethyl alcohol (I.M.S.)
formic acid
glycerol

methyl acetate
monochloroacetic acid
naphthalene
p-nitrotoluene
phenol
phenylacetic acid
pyridine
quinhydrone
sucrose
m-toluidine
trichloroacetic acid
urea

GENERAL

agar-agar
'Biodeminrolit' ion exchange
resin (pH dye indicated)
'De-acidite' FF ion exchange resin
(chloride form, 14–52 mesh) chromatographic grade
charcoal, activated, granular
carbon dioxide, cylinder

sulphur dioxide, siphon or cylinder
mercury
platinum wire
paraffin, liquid, oil bath grade
pumice, granular (12 mesh, say)
sodium flame pencils
starch

INDICATORS

methyl orange (screened)
methyl orange (unscreened)
methyl red
phenolphthalein

potassium chromate (5% solution)
universal indicator (B.T.L.)
starch solution (1% freshly prepared)

(vii) ERRORS, AND ORDER OF ACCURACY

Errors occur whenever a measurement or setting is made, or a reading taken. These errors are usually due to a combination of two factors: the limitations of the instrument used and the judgement of the observer. Sometimes errors are introduced into experimental results because of the nature of the experiment itself (as with heat losses in calorimetry) or because of approximations made in calculation. It is usually possible to estimate the probable error in a particular reading, and this error must be taken into account when calculating the final result if the latter is to have any real meaning.

Consider the measurement of temperature with a 0–50°C thermometer calibrated in 0·1°C. Even with the aid of a lens it will be found exceedingly difficult to estimate the second decimal place to nearer than 0·02°C. Thus a reading of 20·64°C implies that the temperature lies within the 0·02° range from 20·63 to 20·65°C. The possible error is therefore ±0·01°C and the reading should be recorded as **20·64** (**±0·01**)°C. Suppose the temperature is now raised to 32·86°C. This must be recorded as **32·86**(**±0·01**)°C. The temperature rise is 12·22°C but this must be recorded as 12·22 (±0·02)°C, the error of ±0·02°C taking into account of the ±0·01 degrees error **at each end of the range**.

PERCENTAGE ERRORS IN SINGLE READINGS

The percentage error in the reading 20·64°C (recorded as 20·64 (±0·01)°C) is

$$\pm \left[\frac{0·01}{20·64} \times 100 \right] \% \simeq \underline{\pm 0·05\%}$$

PERCENTAGE ERRORS IN SUMS AND DIFFERENCES

The measurement of temperature rise, or fall, involves taking the difference of two thermometer readings. The error in the final result is obtained by **adding** the errors of the two readings whose difference is taken since the error in the larger value might be positive and the error in the smaller value might be negative.

Thus $32·86(\pm 0·01) - 20·64(\pm 0·01) = 12·22(\pm 0·02)$ as above. The percentage error in the final result is

$$\pm \left[\frac{0·02}{12·22} \times 100 \right] \% \simeq \underline{\pm 0·16\%}$$

When two values (such as weights) are added, the errors are added. It is possible that the errors in both values are negative or both positive.

It is well to remember that the percentage error in a result obtained by difference is greater than the percentage error in either of the single measurements, and also that the smaller the difference in the two measurements the greater will be the percentage error in the difference.

PERCENTAGE ERRORS IN PRODUCTS AND QUOTIENTS

Consider the calculation of R from the hypothetical formula

$$R = \frac{4\pi ab}{c(d+e)}$$

where a, b, c, d and e represent readings or measurements taken during an experiment. There will be errors ($\pm \delta a$, $\pm \delta b$, $\pm \delta c$, $\pm \delta d$, etc.) in each of these measurements, and therefore an error ($\pm \delta R$) in the value of R obtained from them.

Taking logarithms and differentiating we have

$$\log R = \log 4 + \log \pi + \log a + \log b - \log c - \log (d+e)$$

$$\frac{\delta R}{R} = + \frac{\delta a}{a} + \frac{\delta b}{b} - \frac{\delta c}{c} - \frac{\delta(d+e)}{d+e}$$

Multiplying through by 100 the terms of the equation are then seen to represent the percentage errors in each of the values concerned. Since the errors can be positive or negative it follows that the percentage error in R is given by

$$\pm 100\,\frac{\delta R}{R} = 100\left[\pm\frac{\delta a}{a}\pm\frac{\delta b}{b}\mp\frac{\delta c}{c}\mp\frac{\delta(d+e)}{(d+e)}\right]$$

and the maximum **percentage error** in R is thus given by **adding** the percentage errors in c and $(d+e)$ to those in a and b.

Now consider the expression

$$X = Y^n$$

Taking logarithms and differentiating we have

$$\log X = n \log Y$$

and

$$\frac{\delta X}{X} = n\,\frac{\delta Y}{Y}$$

Thus it will be seen that the percentage error in a result that is calculated by raising to the power n is n times as great as the percentage error in the reading itself. Great care must be taken to make such readings as accurate as possible if a large error is not to appear in the final result.

Example (Experiment 20, p. 49, Landsberger's method for determining molecular weights)

Boiling point constant for water   = 5·1°c/mole/100 g

Boiling point water from the experiment (T_w) = 98·42 ($\pm$0·01)°c

Weight of urea (W) = 2·121 ($\pm$0·0005) g

Volume of solution ($\simeq$volume of water) (V) = 15·1 ($\pm$0·05) ml

Boiling point of the solution (T_s) = 99·68 ($\pm$0·01)°c

Boiling point elevation $(T_s - T_w)$ = 1·26 ($\pm$0·02)°c

Density of water at the boiling point   = 0·958 g/ml

The molecular weight of the solute is given by

$$\text{M.W.} = \frac{W\times 5{\cdot}1\times 100}{(T_s-T_w)0{\cdot}958V} = \frac{2{\cdot}121\times 5{\cdot}1\times 100}{1{\cdot}26\times 0{\cdot}958\times 15{\cdot}1} = \underline{59{\cdot}35}\ \text{(logs)}$$

The above calculation takes no account of errors.

$$\text{Percentage error in M.W.} = \pm 100\left[\frac{\delta W}{W}+\frac{\delta(T_s-T_w)}{(T_s-T_w)}+\frac{\delta V}{V}\right]\%$$

$$= \pm\left[\frac{0{\cdot}0005\times 100}{2{\cdot}121}+\frac{0{\cdot}02\times 100}{1{\cdot}26}+\frac{0{\cdot}05\times 100}{15{\cdot}1}\right]\%$$

$$\simeq \underline{\pm 2\%}$$

It follows that it is meaningless to record the calculated value of the molecular weight as **59·35** since this implies an order of accuracy of 1 part in 6000, or approximately 0·016%. Similarly, a value given as **59·4** would imply an order of accuracy of 0·16%. We have seen that the probable error in the final result is approximately 2%. Thus the result must be recorded as **59(± 1)**, taking account of the calculated approximate error.

PERCENTAGE ERRORS IN MORE COMPLEX FUNCTIONS

Consider the calculation of A from the hypothetical formula

$$A = \frac{1}{B}\left[4C(D-E)+G(D+F)\right]$$

A is first calculated without considering the errors in the experimental values of B, C, D, E, F and G. The calculation of the percentage error in A is slightly more complicated than in the previous examples since, as we have seen, **in sums and differences** it is **actual errors** which are added while in **products and quotients** it is **percentage (or fractional) errors** which are added.

Taking account of errors ($\pm \delta C$, etc.) the product $4C(D-E)$ becomes

$$4(C \pm \delta C)\left[(D-E) \pm \delta(D-E)\right]$$

which expands to

$$4\left[C(D-E) \pm C\delta(D-E) \pm (D-E)\delta C \pm \delta C\delta(D-E)\right]$$

which is approximately equal to

$$4\left[C(D-E) \pm C\delta(D-E) \pm (D-E)\delta C\right]$$

since $\delta C\delta(D-E)$ is very small.

The actual error in this quantity is

$$4\left[\pm C\delta(D-E) \pm (D-E)\delta C\right]$$

bearing in mind that $\delta(D-E) = \delta D + \delta E$.

Now, referring back to the original formula, it will be seen that the error in the square bracketed factor is

$$\pm \delta[\ ] = \pm 4C(\delta D + \delta E) \pm 4(D-E)\delta C \pm G(\delta D + \delta F) \pm (D+F)\delta G$$

The percentage error in A is thus given by

$$\frac{100\delta A}{A} = \left[\frac{100\delta B}{B} + \frac{100\delta[\ ]}{[\ ]}\right] \%$$

Example (Experiment 8, p. 18: The determination of heats of neutralisation)

In this experiment the water equivalent of the vacuum flask calorimeter is determined first:

Initial temperature of flask and cold water (t_1) $= 15\cdot64\ (\pm 0\cdot01)°c$

Initial temperature of hot water (t_2) $\qquad = 45\cdot22\ (\pm 0\cdot01)°c$

Final temperature of flask and mixture (t_3)   = 28·72 ($\pm$0·01)°C

Volume of cold water (grade 'B' pipette)   = 50·0 ($\pm$0·2) ml

Volume of hot water (grade 'B' pipette)   = 50·0 ($\pm$0·2) ml

The water equivalent (W) of the flask calorimeter is given by

$$W = \frac{50(t_2 - t_3)}{(t_3 - t_1)} - 50 = 63\cdot1 - 50 = \underline{13\cdot1} \text{ g (taking no account of errors)}$$

The percentage error in the first term

$$= \pm \left[\frac{0\cdot2 \times 100}{50} + \frac{0\cdot02 \times 100}{16\cdot5} + \frac{0\cdot02 \times 100}{13\cdot08} \right] \%$$

$$\simeq \pm 0\cdot7\%$$

Therefore the actual error in the first term is

$$\pm \frac{0\cdot7 \times 63\cdot1}{100} \simeq \underline{\pm 0\cdot4 \text{ g}}$$

Therefore the actual error in the water equivalent is

$$\simeq \pm [0\cdot4 + 0\cdot2] = \underline{\pm 0\cdot6 \text{ g}}$$

from which it follows that the value of the water equivalent obtained must be recorded as

$$\underline{13\cdot1 \ (\pm 0\cdot6)\text{g}}$$

The heat of neutralisation is then determined:

Water equivalent of flask calorimeter (W)   = 13·1 ($\pm$0·6) g

Initial temperature of flask and alkali (T_1)   = 15·24 ($\pm$0·01)°C

Initial temperature of acid (T_2)   = 15·82 ($\pm$0·01)°C

Final temperature of flask and mixture (T_3) = 21·52 ($\pm$0·01)°C

Volume of 1·0N HCl (grade 'B' pipette)   = 50·0 ($\pm$0·2) ml

Volume of 1·0N NaOH (grade 'B' pipette)   = 50·0 ($\pm$0·2) ml

50 ml of a 1·0N solution contains 1/20th gram equivalent, therefore the heat of neutralisation is given by

$$\Delta H = 20Q = 20[W(T_1 - T_3) + 50(T_1 - T_3) + 50(T_2 - T_3)] \text{ calories}$$

$$= 20[13\cdot1 \times 6\cdot28 + 50 \times 6\cdot28 + 50 \times 5\cdot70] \text{ calories}$$

$$= \underline{13,630 \text{ calories}} \text{ (taking no account of errors)}$$

The error in Q is given by

$$\pm \delta Q = \pm W\delta(T_1 - T_3) \pm (T_1 - T_3)\delta W \pm 50\delta(T_1 - T_3)$$
$$\pm (T_1 - T_3)0\cdot 2 \pm 50\delta(T_2 - T_3) \pm (T_2 - T_3)0\cdot 2 \text{ calories}$$

$$= \pm 13\cdot 1 \times 0\cdot 02 \pm 6\cdot 28 \times 0\cdot 6 \pm 50 \times 0\cdot 02 \pm 6\cdot 28 \times 0\cdot 2 \pm 50 \times 0\cdot 02 \pm 5\cdot 70 \times 0\cdot 2 \text{ calories}$$

$$= \pm [0\cdot 262 + 3\cdot 768 + 1\cdot 00 + 1\cdot 256 + 1\cdot 00 + 1\cdot 140] \text{ calories}$$

$$= \pm 8\cdot 426 \text{ calories}$$

$$\pm \Delta\delta H = \pm 20\delta Q = \pm 20 \times 8\cdot 426 = \underline{\pm 170 \text{ calories}}$$

Thus the determined heat of neutralisation must be expressed as

$$\underline{\Delta H = 13{,}630 \ (\pm 170) \text{ calories}}$$

The percentage error is

$$\pm \frac{170 \times 100}{13{,}630}\% \backsimeq \underline{\pm 1\cdot 25\%}$$

ERRORS IN THE PLOTTING AND USE OF STRAIGHT LINE GRAPHS

When plotting results on graph paper the use of mere points is rarely justified since either the ordinate or abcissa, or both, will be subject to a certain probable error. Consider the plotting of the point

$$x = 72 \ (\pm 1), \quad y = 17\cdot 6 \ (\pm 0\cdot 5)$$

(much enlarged for clarity)

The errors will have been calculated by one of the methods previously dealt with. The point (72, 17·6) is represented by P, but the errors $\delta x = \pm 1$ and $\delta y = \pm 0\cdot 5$ indicate that this point could lie anywhere within the rectangle ABCD. The horizontal dotted lines represent the probable limits of error in y, and the vertical dotted lines the probable limits of error in x. The results giving this point should thus be plotted as the centre of a small rectangle of sides $2\delta x$, $2\delta y$. (There will be occasions when the error in x or y will be negligible, and in this case the rectangle reduces to a single line.)

When all the points have been plotted in this way, and any unreasonable points rejected, two lines are drawn through the points. These lines are those of maximum and minimum slope obtainable by passing through all of the rectangles surrounding the points to be used. Bisecting the angle produced should yield the 'best' line, and its slope and intercepts may be taken as those desired. The amounts by which the slopes and intercepts of the lines of maximum and minimum slope differ from those of the 'best' line may be taken as the probable limits of error.

An occasion when the above method cannot be used arises when the points are well scattered. It is frequently possible to estimate the 'best' straight line through scattered points, but a better method is to draw several lines (thus obtaining several slopes and intercepts) in the following way:

The points are plotted just as points (without rectangles of probable limits of error) and any point considered to be worthless is rejected. Suppose there remain twelve useful points. Lines are drawn through points 1 and 7, then 2 and 8, 3 and 9, etc., and this gives six slopes and six intercepts. The mean of the six results is taken. The maximum reasonable deviation from the mean may be taken as the probable limit of error, positive and negative.

When results are being analysed graphically it must be remembered that the slope of the graph is given by:

$$\text{Slope} = \frac{y_1 - y_2}{x_1 - x_2} \text{ (with units given)}$$

The slope is **not necessarily** the tangent of the angle formed by the line of the graph and the x axis.

(viii) VAPOUR PRESSURE OF WATER

TABLE OF VALUES (mm OF MERCURY) FROM $10°$C TO $24°$C

$t°$C	p mm	$t°$C	p mm	$t°$C	p mm
10	9·17	15	12·70	20	17·39
11	9·79	16	13·54	21	18·50
12	10·46	17	14·42	22	19.66
13	11·16	18	15·36	23	20·89
14	11·91	19	16·35	24	22·18

GRAPH OF VALUES (mm OF MERCURY) FROM 90°C TO 102°C

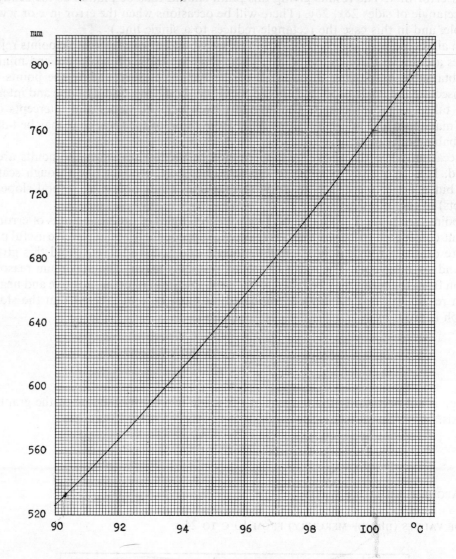

TABLE OF ATOMIC WEIGHTS

Atomic number	Element		Atomic weight (C¹² scale)	Atomic number	Element		Atomic weight (C¹² scale)
1	H	Hydrogen	1·00795	52	Te	Tellurium	127·603
2	He	Helium	4·00282	53	I	Iodine	126·904
3	Li	Lithium	6·93970	54	Xe	Xenon	131·294
4	Be	Beryllium	9·01261	55	Cs	Caesium	132·904
5	B	Boron	10·8195	56	Ba	Barium	137·354
6	C	Carbon	12·0104	57	La	Lanthanum	138·914
7	N	Nitrogen	14·0073	58	Ce	Cerium	140·122
8	O	Oxygen	15·9993	59	Pr	Praseodymium	140·912
9	F	Fluorine	18·9991	60	Nd	Neodymium	144·263
10	Ne	Neon	20·1811	61	Pm	Promethium	(145)
11	Na	Sodium	22·9900	62	Sm	Samarium	150·343
12	Mg	Magnesium	24·3189	63	Eu	Europium	151·993
13	Al	Aluminium	26·9788	64	Gd	Gadolinium	157·253
14	Si	Silicon	28·0087	65	Tb	Terbium	158·932
15	P	Phosphorus	30·9735	66	Dy	Dysprosium	162·503
16	S	Sulphur	32·0646	67	Ho	Holmium	164·932
17	Cl	Chlorine	35·4554	68	Er	Erbium	167·262
18	A	Argon	39·9422	69	Tm	Thulium	168·932
19	K	Potassium	39·0983	70	Yb	Ytterbium	173·032
20	Ca	Calcium	40·0782	71	Lu	Lutetium	174·983
21	Sc	Scandium	44·9580	72	Hf	Hafnium	178·492
22	Ti	Titanium	47·8979	73	Ta	Tantalum	180·942
23	V	Vanadium	50·9478	74	W	Tungsten	183·852
24	Cr	Chromium	52·0077	75	Re	Rhenium	186·211
25	Mn	Manganese	54·9376	76	Os	Osmium	190·191
26	Fe	Iron	55·8475	77	Ir	Iridium	192·191
27	Co	Cobalt	58·9374	78	Pt	Platinum	195·081
28	Ni	Nickel	58·7074	79	Au	Gold	196·991
29	Cu	Copper	63·5371	80	Hg	Mercury	200·601
30	Zn	Zinc	65·3771	81	Tl	Thallium	204·381
31	Ga	Gallium	69·7170	82	Pb	Lead	207·201
32	Ge	Germanium	72·5968	83	Bi	Bismuth	208·991
33	As	Arsenic	74·9067	84	Po	Polonium	209·990
34	Se	Selenium	78·9565	85	At	Astatine	(210)
35	Br	Bromine	79·9125	86	Rn	Radon	221·990
36	Kr	Krypton	83·7963	87	Fr	Francium	(223)
37	Rb	Rubidium	85·4763	88	Ra	Radium	226·040
38	Sr	Strontium	87·6262	89	Ac	Actinium	(227)
39	Y	Yttrium	88·9161	90	Th	Thorium	232·040
40	Zr	Zirconium	91·2160	91	Pa	Protoactinium	230·990
41	Nb	Niobium	92·9060	92	U	Uranium	238·059
42	Mo	Molybdenum	95·9458	93	Np	Neptunium	(237)
43	Tc	Technetium	(99)	94	Pu	Plutonium	(242)
44	Ru	Ruthenium	101·095	95	Am	Americium	(243)
45	Rh	Rhodium	102·905	96	Cm	Curium	(243)
46	Pd	Palladium	106·395	97	Bk	Berkelium	(245)
47	Ag	Silver	107·875	98	Cf	Californium	(246)
48	Cd	Cadmium	112·405	99	Es	Einsteinium	
49	In	Indium	114·815	100	Fm	Fermium	
50	Sn	Tin	118·694	101	Md	Mendelevium	255·990
51	Sb	Antimony	121·754	102	No	Nobelium	
				103	Lw	Lawrencium	

32   **RECIPROCALS**

SUBTRACT.

	0	1	2	3	4	5	6	7	8	9	1	2	3	4	5	6	7	8	9
10	10000	9901	9804	9709	9615	9524	9434	9346	9259	9174	9	18	28	37	46	55	64	73	83
11	9091	9009	8929	8850	8772	8696	8621	8547	8475	8403	8	15	23	31	38	46	54	61	69
12	8333	8264	8197	8130	8065	8000	7937	7874	7813	7752	6	13	19	26	32	39	45	52	58
13	7692	7634	7576	7519	7463	7407	7353	7299	7246	7194	5	11	17	22	28	33	39	44	50
14	7143	7092	7042	6993	6944	6897	6849	6803	6757	6711	5	10	14	19	24	29	34	38	43
15	6667	6623	6579	6536	6494	6452	6410	6369	6329	6289	4	8	13	17	21	25	29	34	38
16	6250	6211	6173	6135	6098	6061	6024	5988	5952	5917	4	7	11	15	18	22	26	29	33
17	5882	5848	5814	5780	5747	5714	5682	5650	5618	5587	3	7	10	13	16	20	23	26	29
18	5556	5525	5495	5464	5435	5405	5376	5348	5319	5291	3	6	9	12	15	18	20	23	26
19	5263	5236	5208	5181	5155	5128	5102	5076	5051	5025	3	5	8	11	13	16	18	21	24
20	5000	4975	4950	4926	4902	4878	4854	4831	4808	4785	2	5	7	10	12	14	17	19	21
21	4762	4739	4717	4695	4673	4651	4630	4608	4587	4566	2	4	7	9	11	13	15	17	19
22	4545	4525	4505	4484	4464	4444	4425	4405	4386	4367	2	4	6	8	10	12	14	16	18
23	4348	4329	4310	4292	4274	4255	4237	4219	4202	4184	2	4	5	7	9	11	13	15	16
24	4167	4149	4132	4115	4098	4082	4065	4049	4032	4016	2	3	5	7	8	10	12	13	15
25	4000	3984	3968	3953	3937	3922	3906	3891	3876	3861	2	3	5	6	8	9	11	12	14
26	3846	3831	3817	3802	3788	3774	3759	3745	3731	3717	1	3	4	6	7	9	10	11	13
27	3704	3690	3676	3663	3650	3636	3623	3610	3597	3584	1	3	4	5	7	8	9	11	12
28	3571	3559	3546	3534	3521	3509	3497	3484	3472	3460	1	2	4	5	6	7	9	10	11
29	3448	3436	3425	3413	3401	3390	3378	3367	3356	3344	1	2	3	5	6	7	8	9	10
30	3333	3322	3311	3300	3289	3279	3268	3257	3247	3236	1	2	3	4	5	6	8	9	10
31	3226	3215	3205	3195	3185	3175	3165	3155	3145	3135	1	2	3	4	5	6	7	8	9
32	3125	3115	3106	3096	3086	3077	3067	3058	3049	3040	1	2	3	4	5	6	7	8	9
33	3030	3021	3012	3003	2994	2985	2976	2967	2959	2950	1	2	3	4	4	5	6	7	8
34	2941	2933	2924	2915	2907	2899	2890	2882	2874	2865	1	2	3	3	4	5	6	7	8
35	2857	2849	2841	2833	2825	2817	2809	2801	2793	2786	1	2	2	3	4	5	6	6	7
36	2778	2770	2762	2755	2747	2740	2732	2725	2717	2710	1	2	2	3	4	5	5	6	7
37	2703	2695	2688	2681	2674	2667	2660	2653	2646	2639	1	1	2	3	4	4	5	6	6
38	2632	2625	2618	2611	2604	2597	2591	2584	2577	2571	1	1	2	3	3	4	5	5	6
39	2564	2558	2551	2545	2538	2532	2525	2519	2513	2506	1	1	2	3	3	4	4	5	6
40	2500	2494	2488	2481	2475	2469	2463	2457	2451	2445	1	1	2	2	3	4	4	5	5
41	2439	2433	2427	2421	2415	2410	2404	2398	2392	2387	1	1	2	2	3	3	4	5	5
42	2381	2375	2370	2364	2358	2353	2347	2342	2336	2331	1	1	2	2	3	3	4	4	5
43	2326	2320	2315	2309	2304	2299	2294	2288	2283	2278	1	1	2	2	3	3	4	4	5
44	2273	2268	2262	2257	2252	2247	2242	2237	2232	2227	1	1	2	2	3	3	4	4	5
45	2222	2217	2212	2208	2203	2198	2193	2188	2183	2179	0	1	1	2	2	3	3	4	4
46	2174	2169	2165	2160	2155	2151	2146	2141	2137	2132	0	1	1	2	2	3	3	4	4
47	2128	2123	2119	2114	2110	2105	2101	2096	2092	2088	0	1	1	2	2	3	3	3	4
48	2083	2079	2075	2070	2066	2062	2058	2053	2049	2045	0	1	1	2	2	2	3	3	4
49	2041	2037	2033	2028	2024	2020	2016	2012	2008	2004	0	1	1	2	2	2	3	3	4
50	2000	1996	1992	1988	1984	1980	1976	1972	1969	1965	0	1	1	2	2	2	3	3	3
51	1961	1957	1953	1949	1946	1942	1938	1934	1931	1927	0	1	1	2	2	2	3	3	3
52	1923	1919	1916	1912	1908	1905	1901	1898	1894	1890	0	1	1	1	2	2	3	3	3
53	1887	1883	1880	1876	1873	1869	1866	1862	1859	1855	0	1	1	1	2	2	2	3	3
54	1852	1848	1845	1842	1838	1835	1832	1828	1825	1821	0	1	1	1	2	2	2	3	3

SUBTRACT.

Find the position of the decimal point by inspection.

We are indebted to Messrs. G. Bell & Sons for permission to reproduce the tables on this and succeeding pages.

RECIPROCALS 33

SUBTRACT.

	0	1	2	3	4	5	6	7	8	9	1	2	3	4	5	6	7	8	9
55	1818	1815	1812	1808	1805	1802	1799	1795	1792	1789	0	1	1	1	2	2	2	3	3
56	1786	1783	1779	1776	1773	1770	1767	1764	1761	1757	0	1	1	1	2	2	2	3	3
57	1754	1751	1748	1745	1742	1739	1736	1733	1730	1727	0	1	1	1	2	2	2	2	3
58	1724	1721	1718	1715	1712	1709	1706	1704	1701	1698	0	1	1	1	1	2	2	2	3
59	1695	1692	1689	1686	1684	1681	1678	1675	1672	1669	0	1	1	1	1	2	2	2	3
60	1667	1664	1661	1658	1656	1653	1650	1647	1645	1642	0	1	1	1	1	2	2	2	2
61	1639	1637	1634	1631	1629	1626	1623	1621	1618	1616	0	1	1	1	1	2	2	2	2
62	1613	1610	1608	1605	1603	1600	1597	1595	1592	1590	0	1	1	1	1	2	2	2	2
63	1587	1585	1582	1580	1577	1575	1572	1570	1567	1565	0	0	1	1	1	1	2	2	2
64	1563	1560	1558	1555	1553	1550	1548	1546	1543	1541	0	0	1	1	1	1	2	2	2
65	1538	1536	1534	1531	1529	1527	1524	1522	1520	1517	0	0	1	1	1	1	2	2	2
66	1515	1513	1511	1508	1506	1504	1502	1499	1497	1495	0	0	1	1	1	1	2	2	2
67	1493	1490	1488	1486	1484	1481	1479	1477	1475	1473	0	0	1	1	1	1	2	2	2
68	1471	1468	1466	1464	1462	1460	1458	1456	1453	1451	0	0	1	1	1	1	1	2	2
69	1449	1447	1445	1443	1441	1439	1437	1435	1433	1431	0	0	1	1	1	1	1	2	2
70	1429	1427	1425	1422	1420	1418	1416	1414	1412	1410	0	0	1	1	1	1	1	2	2
71	1408	1406	1404	1403	1401	1399	1397	1395	1393	1391	0	0	1	1	1	1	1	2	2
72	1389	1387	1385	1383	1381	1379	1377	1376	1374	1372	0	0	1	1	1	1	1	2	2
73	1370	1368	1366	1364	1362	1361	1359	1357	1355	1353	0	0	1	1	1	1	1	2	2
74	1351	1350	1348	1346	1344	1342	1340	1339	1337	1335	0	0	1	1	1	1	1	1	2
75	1333	1332	1330	1328	1326	1325	1323	1321	1319	1318	0	0	1	1	1	1	1	1	2
76	1316	1314	1312	1311	1309	1307	1305	1304	1302	1300	0	0	1	1	1	1	1	1	2
77	1299	1297	1295	1294	1292	1290	1289	1287	1285	1284	0	0	0	1	1	1	1	1	1
78	1282	1280	1279	1277	1276	1274	1272	1271	1269	1267	0	0	0	1	1	1	1	1	1
79	1266	1264	1263	1261	1259	1258	1256	1255	1253	1252	0	0	0	1	1	1	1	1	1
80	1250	1248	1247	1245	1244	1242	1241	1239	1238	1236	0	0	0	1	1	1	1	1	1
81	1235	1233	1232	1230	1229	1227	1225	1224	1222	1221	0	0	0	1	1	1	1	1	1
82	1220	1218	1217	1215	1214	1212	1211	1209	1208	1206	0	0	0	1	1	1	1	1	1
83	1205	1203	1202	1200	1199	1198	1196	1195	1193	1192	0	0	0	1	1	1	1	1	1
84	1190	1189	1188	1186	1185	1183	1182	1181	1179	1178	0	0	0	1	1	1	1	1	1
85	1176	1175	1174	1172	1171	1170	1168	1167	1166	1164	0	0	0	1	1	1	1	1	1
86	1163	1161	1160	1159	1157	1156	1155	1153	1152	1151	0	0	0	1	1	1	1	1	1
87	1149	1148	1147	1145	1144	1143	1142	1140	1139	1138	0	0	0	1	1	1	1	1	1
88	1136	1135	1134	1133	1131	1130	1129	1127	1126	1125	0	0	0	1	1	1	1	1	1
89	1124	1122	1121	1120	1119	1117	1116	1115	1114	1112	0	0	0	1	1	1	1	1	1
90	1111	1110	1109	1107	1106	1105	1104	1103	1101	1100	0	0	0	0	1	1	1	1	1
91	1099	1098	1096	1095	1094	1093	1092	1091	1089	1088	0	0	0	0	1	1	1	1	1
92	1087	1086	1085	1083	1082	1081	1080	1079	1078	1076	0	0	0	0	1	1	1	1	1
93	1075	1074	1073	1072	1071	1070	1068	1067	1066	1065	0	0	0	0	1	1	1	1	1
94	1064	1063	1062	1060	1059	1058	1057	1056	1055	1054	0	0	0	0	1	1	1	1	1
95	1053	1052	1050	1049	1048	1047	1046	1045	1044	1043	0	0	0	0	1	1	1	1	1
96	1042	1041	1040	1038	1037	1036	1035	1034	1033	1032	0	0	0	0	1	1	1	1	1
97	1031	1030	1029	1028	1027	1026	1025	1024	1022	1021	0	0	0	0	1	1	1	1	1
98	1020	1019	1018	1017	1016	1015	1014	1013	1012	1011	0	0	0	0	0	1	1	1	1
99	1010	1009	1008	1007	1006	1005	1004	1003	1002	1001	0	0	0	0	0	1	1	1	1

SUBTRACT.

Find the position of the decimal point by inspection.

LOGARITHMS

	0	1	2	3	4	5	6	7	8	9	1	2	3	4	5	6	7	8	9
10	·0000	0043	0086	0128	0170	0212	0253	0294	0334	0374	4	8	12	17	21	25	29	33	37
11	·0414	0453	0492	0531	0569	0607	0645	0682	0719	0755	4	8	11	15	19	23	26	30	34
12	·0792	0828	0864	0899	0934	0969	1004	1038	1072	1106	3	7	10	14	17	21	24	28	31
13	·1139	1173	1206	1239	1271	1303	1335	1367	1399	1430	3	6	10	13	16	19	23	26	29
14	·1461	1492	1523	1553	1584	1614	1644	1673	1703	1732	3	6	9	12	15	18	21	24	27
15	·1761	1790	1818	1847	1875	1903	1931	1959	1987	2014	3	6	8	11	14	17	20	22	25
16	·2041	2068	2095	2122	2148	2175	2201	2227	2253	2279	3	5	8	11	13	16	18	21	24
17	·2304	2330	2355	2380	2405	2430	2455	2480	2504	2529	2	5	7	·10	12	15	17	20	22
18	·2553	2577	2601	2625	2648	2672	2695	2718	2742	2765	2	5	7	9	12	14	16	19	21
19	·2788	2810	2833	2856	2878	2900	2923	2945	2967	2989	2	4	7	9	11	13	16	18	20
20	·3010	3032	3054	3075	3096	3118	3139	3160	3181	3201	2	4	6	8	11	13	15	17	19
21	·3222	3243	3263	3284	3304	3324	3345	3365	3385	3404	2	4	6	8	10	12	14	16	18
22	·3424	3444	3464	3483	3502	3522	3541	3560	3579	3598	2	4	6	8	10	12	14	15	17
23	·3617	3636	3655	3674	3692	3711	3729	3747	3766	3784	2	4	6	7	9	11	13	15	17
24	·3802	3820	3838	3856	3874	3892	3909	3927	3945	3962	2	4	5	7	9	11	12	14	16
25	·3979	3997	4014	4031	4048	4065	4082	4099	4116	4133	2	3	5	7	9	10	12	14	15
26	·4150	4166	4183	4200	4216	4232	4249	4265	4281	4298	2	3	5	7	8	10	11	13	15
27	·4314	4330	4346	4362	4378	4393	4409	4425	4440	4456	2	3	5	6	8	9	11	13	14
28	·4472	4487	4502	4518	4533	4548	4564	4579	4594	4609	2	3	5	6	8	9	11	12	14
29	·4624	4639	4654	4669	4683	4698	4713	4728	4742	4757	1	3	4	6	7	9	10	12	13
30	·4771	4786	4800	4814	4829	4843	4857	4871	4886	4900	1	3	4	6	7	9	10	11	13
31	·4914	4928	4942	4955	4969	4983	4997	5011	5024	5038	1	3	4	6	7	8	10	11	12
32	·5051	5065	5079	5092	5105	5119	5132	5145	5159	5172	1	3	4	5	7	8	9	11	12
33	·5185	5198	5211	5224	5237	5250	5263	5276	5289	5302	1	3	4	5	6	8	9	10	12
34	·5315	5328	5340	5353	5366	5378	5391	5403	5416	5428	1	3	4	5	6	8	9	10	11
35	·5441	5453	5465	5478	5490	5502	5514	5527	5539	5551	1	2	4	5	6	7	9	10	11
36	·5563	5575	5587	5599	5611	5623	5635	5647	5658	5670	1	2	4	5	6	7	8	10	11
37	·5682	5694	5705	5717	5729	5740	5752	5763	5775	5786	1	2	3	5	6	7	8	9	10
38	·5798	5809	5821	5832	5843	5855	5866	5877	5888	5899	1	2	3	5	6	7	8	9	10
39	·5911	5922	5933	5944	5955	5966	5977	5988	5999	6010	1	2	3	4	5	7	8	9	10
40	·6021	6031	6042	6053	6064	6075	6085	6096	6107	6117	1	2	3	4	5	6	8	9	10
41	·6128	6138	6149	6160	6170	6180	6191	6201	6212	6222	1	2	3	4	5	6	7	8	9
42	·6232	6243	6253	6263	6274	6284	6294	6304	6314	6325	1	2	3	4	5	6	7	8	9
43	·6335	6345	6355	6365	6375	6385	6395	6405	6415	6425	1	2	3	4	5	6	7	8	9
44	·6435	6444	6454	6464	6474	6484	6493	6503	6513	6522	1	2	3	4	5	6	7	8	9
45	·6532	6542	6551	6561	6571	6580	6590	6599	6609	6618	1	2	3	4	5	6	7	8	9
46	·6628	6637	6646	6656	6665	6675	6684	6693	6702	6712	1	2	3	4	5	6	7	7	8
47	·6721	6730	6739	6749	6758	6767	6776	6785	6794	6803	1	2	3	4	5	5	6	7	8
48	·6812	6821	6830	6839	6848	6857	6866	6875	6884	6893	1	2	3	4	4	5	6	7	8
49	·6902	6911	6920	6928	6937	6946	6955	6964	6972	6981	1	2	3	4	4	5	6	7	8
50	·6990	6998	7007	7016	7024	7033	7042	7050	7059	7067	1	2	3	3	4	5	6	7	8
51	·7076	7084	7093	7101	7110	7118	7126	7135	7143	7152	1	2	3	3	4	5	6	7	8
52	·7160	7168	7177	7185	7193	7202	7210	7218	7226	7235	1	2	2	3	4	5	6	7	7
53	·7243	7251	7259	7267	7275	7284	7292	7300	7308	7316	1	2	2	3	4	5	6	6	7
54	·7324	7332	7340	7348	7356	7364	7372	7380	7388	7396	1	2	2	3	4	5	6	6	7

LOGARITHMS

	0	1	2	3	4	5	6	7	8	9	1	2	3	4	5	6	7	8	9
55	·7404	7412	7419	7427	7435	7443	7451	7459	7466	7474	1	2	2	3	4	5	5	6	7
56	·7482	7490	7497	7505	7513	7520	7528	7536	7543	7551	1	2	2	3	4	5	5	6	7
57	·7559	7566	7574	7582	7589	7597	7604	7612	7619	7627	1	2	2	3	4	5	5	6	7
58	·7634	7642	7649	7657	7664	7672	7679	7686	7694	7701	1	1	2	3	4	4	5	6	7
59	·7709	7716	7723	7731	7738	7745	7752	7760	7767	7774	1	1	2	3	4	4	5	6	7
60	·7782	7789	7796	7803	7810	7818	7825	7832	7839	7846	1	1	2	3	4	4	5	6	6
61	·7853	7860	7868	7875	7882	7889	7896	7903	7910	7917	1	1	2	3	4	4	5	6	6
62	·7924	7931	7938	7945	7952	7959	7966	7973	7980	7987	1	1	2	3	3	4	5	6	6
63	·7993	8000	8007	8014	8021	8028	8035	8041	8048	8055	1	1	2	3	3	4	5	5	6
64	·8062	8069	8075	8082	8089	8096	8102	8109	8116	8122	1	1	2	3	3	4	5	5	6
65	·8129	8136	8142	8149	8156	8162	8169	8176	8182	8189	1	1	2	3	3	4	5	5	6
66	·8195	8202	8209	8215	8222	8228	8235	8241	8248	8254	1	1	2	3	3	4	5	5	6
67	·8261	8267	8274	8280	8287	8293	8299	8306	8312	8319	1	1	2	3	3	4	5	5	6
68	·8325	8331	8338	8344	8351	8357	8363	8370	8376	8382	1	1	2	3	3	4	4	5	6
69	·8388	8395	8401	8407	8414	8420	8426	8432	8439	8445	1	1	2	2	3	4	4	5	6
70	·8451	8457	8463	8470	8476	8482	8488	8494	8500	8506	1	1	2	2	3	4	4	5	6
71	·8513	8519	8525	8531	8537	8543	8549	8555	8561	8567	1	1	2	2	3	4	4	5	5
72	·8573	8579	8585	8591	8597	8603	8609	8615	8621	8627	1	1	2	2	3	4	4	5	5
73	·8633	8639	8645	8651	8657	8663	8669	8675	8681	8686	1	1	2	2	3	4	4	5	5
74	·8692	8698	8704	8710	8716	8722	8727	8733	8739	8745	1	1	2	2	3	4	4	5	5
75	·8751	8756	8762	8768	8774	8779	8785	8791	8797	8802	1	1	2	2	3	3	4	5	5
76	·8808	8814	8820	8825	8831	8837	8842	8848	8854	8859	1	1	2	2	3	3	4	5	5
77	·8865	8871	8876	8882	8887	8893	8899	8904	8910	8915	1	1	2	2	3	3	4	4	5
78	·8921	8927	8932	8938	8943	8949	8954	8960	8965	8971	1	1	2	2	3	3	4	4	5
79	·8976	8982	8987	8993	8998	9004	9009	9015	9020	9025	1	1	2	2	3	3	4	4	5
80	·9031	9036	9042	9047	9053	9058	9063	9069	9074	9079	1	1	2	2	3	3	4	4	5
81	·9085	9090	9096	9101	9106	9112	9117	9122	9128	9133	1	1	2	2	3	3	4	4	5
82	·9138	9143	9149	9154	9159	9165	9170	9175	9180	9186	1	1	2	2	3	3	4	4	5
83	·9191	9196	9201	9206	9212	9217	9222	9227	9232	9238	1	1	2	2	3	3	4	4	5
84	·9243	9248	9253	9258	9263	9269	9274	9279	9284	9289	1	1	2	2	3	3	4	4	5
85	·9294	9299	9304	9309	9315	9320	9325	9330	9335	9340	1	1	2	2	3	3	4	4	5
86	·9345	9350	9355	9360	9365	9370	9375	9380	9385	9390	1	1	2	2	3	3	4	4	5
87	·9395	9400	9405	9410	9415	9420	9425	9430	9435	9440	0	1	1	2	2	3	3	4	4
88	·9445	9450	9455	9460	9465	9469	9474	9479	9484	9489	0	1	1	2	2	3	3	4	4
89	·9494	9499	9504	9509	9513	9518	9523	9528	9533	9538	0	1	1	2	2	3	3	4	4
90	·9542	9547	9552	9557	9562	9566	9571	9576	9581	9586	0	1	1	2	2	3	3	4	4
91	·9590	9595	9600	9605	9609	9614	9619	9624	9628	9633	0	1	1	2	2	3	3	4	4
92	·9638	9643	9647	9652	9657	9661	9666	9671	9675	9680	0	1	1	2	2	3	3	4	4
93	·9685	9689	9694	9699	9703	9708	9713	9717	9722	9727	0	1	1	2	2	3	3	4	4
94	·9731	9736	9741	9745	9750	9754	9759	9763	9768	9773	0	1	1	2	2	3	3	4	4
95	·9777	9782	9786	9791	9795	9800	9805	9809	9814	9818	0	1	1	2	2	3	3	4	4
96	·9823	9827	9832	9836	9841	9845	9850	9854	9859	9863	0	1	1	2	2	3	3	4	4
97	·9868	9872	9877	9881	9886	9890	9894	9899	9903	9908	0	1	1	2	2	3	3	4	4
98	·9912	9917	9921	9926	9930	9934	9939	9943	9948	9952	0	1	1	2	2	3	3	4	4
99	·9956	9961	9965	9969	9974	9978	9983	9987	9991	9996	0	1	1	2	2	3	3	3	4

ANTI-LOGARITHMS

	0	1	2	3	4	5	6	7	8	9	1	2	3	4	5	6	7	8	9
·00	1000	1002	1005	1007	1009	1012	1014	1016	1019	1021	0	0	1	1	1	1	2	2	2
·01	1023	1026	1028	1030	1033	1035	1038	1040	1042	1045	0	0	1	1	1	1	2	2	2
·02	1047	1050	1052	1054	1057	1059	1062	1064	1067	1069	0	0	1	1	1	1	2	2	2
·03	1072	1074	1076	1079	1081	1084	1086	1089	1091	1094	0	0	1	1	1	1	2	2	2
·04	1096	1099	1102	1104	1107	1109	1112	1114	1117	1119	0	1	1	1	1	2	2	2	2
·05	1122	1125	1127	1130	1132	1135	1138	1140	1143	1146	0	1	1	1	1	2	2	2	2
·06	1148	1151	1153	1156	1159	1161	1164	1167	1169	1172	0	1	1	1	1	2	2	2	2
·07	1175	1178	1180	1183	1186	1189	1191	1194	1197	1199	0	1	1	1	1	2	2	2	2
·08	1202	1205	1208	1211	1213	1216	1219	1222	1225	1227	0	1	1	1	1	2	2	2	3
·09	1230	1233	1236	1239	1242	1245	1247	1250	1253	1256	0	1	1	1	1	2	2	2	3
·10	1259	1262	1265	1268	1271	1274	1276	1279	1282	1285	0	1	1	1	1	2	2	2	3
·11	1288	1291	1294	1297	1300	1303	1306	1309	1312	1315	0	1	1	1	2	2	2	2	3
·12	1318	1321	1324	1327	1330	1334	1337	1340	1343	1346	0	1	1	1	2	2	2	2	3
·13	1349	1352	1355	1358	1361	1365	1368	1371	1374	1377	0	1	1	1	2	2	2	3	3
·14	1380	1384	1387	1390	1393	1396	1400	1403	1406	1409	0	1	1	1	2	2	2	3	3
·15	1413	1416	1419	1422	1426	1429	1432	1435	1439	1442	0	1	1	1	2	2	2	3	3
·16	1445	1449	1452	1455	1459	1462	1466	1469	1472	1476	0	1	1	1	2	2	2	3	3
·17	1479	1483	1486	1489	1493	1496	1500	1503	1507	1510	0	1	1	1	2	2	2	3	3
·18	1514	1517	1521	1524	1528	1531	1535	1538	1542	1545	0	1	1	1	2	2	2	3	3
·19	1549	1552	1556	1560	1563	1567	1570	1574	1578	1581	0	1	1	1	2	2	3	3	3
·20	1585	1589	1592	1596	1600	1603	1607	1611	1614	1618	0	1	1	1	2	2	3	3	3
·21	1622	1626	1629	1633	1637	1641	1644	1648	1652	1656	0	1	1	2	2	2	3	3	3
·22	1660	1663	1667	1671	1675	1679	1683	1687	1690	1694	0	1	1	2	2	2	3	3	3
·23	1698	1702	1706	1710	1714	1718	1722	1726	1730	1734	0	1	1	2	2	2	3	3	4
·24	1738	1742	1746	1750	1754	1758	1762	1766	1770	1774	0	1	1	2	2	2	3	3	4
·25	1778	1782	1786	1791	1795	1799	1803	1807	1811	1816	0	1	1	2	2	2	3	3	4
·26	1820	1824	1828	1832	1837	1841	1845	1849	1854	1858	0	1	1	2	2	3	3	3	4
·27	1862	1866	1871	1875	1879	1884	1888	1892	1897	1901	0	1	1	2	2	3	3	3	4
·28	1905	1910	1914	1919	1923	1928	1932	1936	1941	1945	0	1	1	2	2	3	3	4	4
·29	1950	1954	1959	1963	1968	1972	1977	1982	1986	1991	0	1	1	2	2	3	3	4	4
·30	1995	2000	2004	2009	2014	2018	2023	2028	2032	2037	0	1	1	2	2	3	3	4	4
·31	2042	2046	2051	2056	2061	2065	2070	2075	2080	2084	0	1	1	2	2	3	3	4	4
·32	2089	2094	2099	2104	2109	2113	2118	2123	2128	2133	0	1	1	2	2	3	3	4	4
·33	2138	2143	2148	2153	2158	2163	2168	2173	2178	2183	0	1	1	2	2	3	3	4	4
·34	2188	2193	2198	2203	2208	2213	2218	2223	2228	2234	1	1	2	2	3	3	4	4	5
·35	2239	2244	2249	2254	2259	2265	2270	2275	2280	2286	1	1	2	2	3	3	4	4	5
·36	2291	2296	2301	2307	2312	2317	2323	2328	2333	2339	1	1	2	2	3	3	4	4	5
·37	2344	2350	2355	2360	2366	2371	2377	2382	2388	2393	1	1	2	2	3	3	4	4	5
·38	2399	2404	2410	2415	2421	2427	2432	2438	2443	2449	1	1	2	2	3	3	4	4	5
·39	2455	2460	2466	2472	2477	2483	2489	2495	2500	2506	1	1	2	2	3	3	4	5	5
·40	2512	2518	2523	2529	2535	2541	2547	2553	2559	2564	1	1	2	2	3	4	4	5	5
·41	2570	2576	2582	2588	2594	2600	2606	2612	2618	2524	1	1	2	2	3	4	4	5	5
·42	2630	2636	2642	2649	2655	2661	2667	2673	2679	2685	1	1	2	2	3	4	4	5	6
·43	2692	2698	2704	2710	2716	2723	2729	2735	2742	2748	1	1	2	3	3	4	4	5	6
·44	2754	2761	2767	2773	2780	2786	2793	2799	2805	2812	1	1	2	3	3	4	4	5	6
·45	2818	2825	2831	2838	2844	2851	2858	2864	2871	2877	1	1	2	3	3	4	5	5	6
·46	2884	2891	2897	2904	2911	2917	2924	2931	2938	2944	1	1	2	3	3	4	5	5	6
·47	2951	2958	2965	2972	2979	2985	2992	2999	3006	3013	1	1	2	3	3	4	5	5	6
·48	3020	3027	3034	3041	3048	3055	3062	3069	3076	3083	1	1	2	3	4	4	5	6	6
·49	3090	3097	3105	3112	3119	3126	3133	3141	3148	3155	1	1	2	3	4	4	5	6	6

ANTI-LOGARITHMS

	0	1	2	3	4	5	6	7	8	9	1	2	3	4	5	6	7	8	9
·50	3162	3170	3177	3184	3192	3199	3206	3214	3221	3228	1	1	2	3	4	4	5	6	7
·51	3236	3243	3251	3258	3266	3273	3281	3289	3296	3304	1	2	2	3	4	5	5	6	7
·52	3311	3319	3327	3334	3342	3350	3357	3365	3373	3381	1	2	2	3	4	5	5	6	7
·53	3388	3396	3404	3412	3420	3428	3436	3443	3451	3459	1	2	2	3	4	5	6	6	7
·54	3467	3475	3483	3491	3499	3508	3516	3524	3532	3540	1	2	2	3	4	5	6	6	7
·55	3548	3556	3565	3573	3581	3589	3597	3606	3614	3622	1	2	2	3	4	5	6	7	7
·56	3631	3639	3648	3656	3664	3673	3681	3690	3698	3707	1	2	3	3	4	5	6	7	8
·57	3715	3724	3733	3741	3750	3758	3767	3776	3784	3793	1	2	3	3	4	5	6	7	8
·58	3802	3811	3819	3828	3837	3846	3855	3864	3873	3882	1	2	3	4	4	5	6	7	8
·59	3890	3899	3908	3917	3926	3936	3945	3954	3963	3972	1	2	3	4	5	5	6	7	8
·60	3981	3990	3999	4009	4018	4027	4036	4046	4055	4064	1	2	3	4	5	6	6	7	8
·61	4074	4083	4093	4102	4111	4121	4130	4140	4150	4159	1	2	3	4	5	6	7	8	9
·62	4169	4178	4188	4198	4207	4217	4227	4236	4246	4256	1	2	3	4	5	6	7	8	9
·63	4266	4276	4285	4295	4305	4315	4325	4335	4345	4355	1	2	3	4	5	6	7	8	9
·64	4365	4375	4385	4395	4406	4416	4426	4436	4446	4457	1	2	3	4	5	6	7	8	9
·65	4467	4477	4487	4498	4508	4519	4529	4539	4550	4560	1	2	3	4	5	6	7	8	9
·66	4571	4581	4592	4603	4613	4624	4634	4645	4656	4667	1	2	3	4	5	6	7	9	10
·67	4677	4688	4699	4710	4721	4732	4742	4753	4764	4775	1	2	3	4	5	7	8	9	10
·68	4786	4797	4808	4819	4831	4842	4853	4864	4875	4887	1	2	3	4	6	7	8	9	10
·69	4898	4909	4920	4932	4943	4955	4966	4977	4989	5000	1	2	3	5	6	7	8	9	10
·70	5012	5023	5035	5047	5058	5070	5082	5093	5105	5117	1	2	4	5	6	7	8	9	11
·71	5129	5140	5152	5164	5176	5188	5200	5212	5224	5236	1	2	4	5	6	7	8	10	11
·72	5248	5260	5272	5284	5297	5309	5321	5333	5346	5358	1	2	4	5	6	7	9	10	11
·73	5370	5383	5395	5408	5420	5433	5445	5458	5470	5483	1	3	4	5	6	8	9	10	11
·74	5495	5508	5521	5534	5546	5559	5572	5585	5598	5610	1	3	4	5	6	8	9	10	12
·75	5623	5636	5649	5662	5675	5689	5702	5715	5728	5741	1	3	4	5	7	8	9	10	12
·76	5754	5768	5781	5794	5808	5821	5834	5848	5861	5875	1	3	4	5	7	8	9	11	12
·77	5888	5902	5916	5929	5943	5957	5970	5984	5998	6012	1	3	4	5	7	8	10	11	12
·78	6026	6039	6053	6067	6081	6095	6109	6124	6138	6152	1	3	4	6	7	8	10	11	13
·79	6166	6180	6194	6209	6223	6237	6252	6266	6281	6295	1	3	4	6	7	9	10	11	13
·80	6310	6324	6339	6353	6368	6383	6397	6412	6427	6442	1	3	4	6	7	9	10	12	13
·81	6457	6471	6486	6501	6516	6531	6546	6561	6577	6592	2	3	5	6	8	9	11	12	14
·82	6607	6622	6637	6653	6668	6683	6699	6714	6730	6745	2	3	5	6	8	9	11	12	14
·83	6761	6776	6792	6808	6823	6839	6855	6871	6887	6902	2	3	5	6	8	9	11	13	14
·84	6918	6934	6950	6966	6982	6998	7015	7031	7047	7063	2	3	5	6	8	10	11	13	15
·85	7079	7096	7112	7129	7145	7161	7178	7194	7211	7228	2	3	5	7	8	10	12	13	15
·86	7244	7261	7278	7295	7311	7328	7345	7362	7379	7396	2	3	5	7	8	10	12	13	15
·87	7413	7430	7447	7464	7482	7499	7516	7534	7551	7568	2	3	5	7	9	10	12	14	16
·88	7586	7603	7621	7638	7656	7674	7691	7709	7727	7745	2	4	5	7	9	11	12	14	16
·89	7762	7780	7798	7816	7834	7852	7870	7889	7907	7925	2	4	5	7	9	11	13	14	16
·90	7943	7962	7980	7998	8017	8035	8054	8072	8091	8110	2	4	6	7	9	11	13	15	17
·91	8128	8147	8166	8185	8204	8222	8241	8260	8279	8299	2	4	6	8	9	11	13	15	17
·92	8318	8337	8356	8375	8395	8414	8433	8453	8472	8492	2	4	6	8	10	12	14	15	17
·93	8511	8531	8551	8570	8590	8610	8630	8650	8670	8690	2	4	6	8	10	12	14	16	18
·94	8710	8730	8750	8770	8790	8810	8831	8851	8872	8892	2	4	6	8	10	12	14	16	18
·95	8913	8933	8954	8974	8995	9016	9036	9057	9078	9099	2	4	6	8	10	12	15	17	19
·96	9120	9141	9162	9183	9204	9226	9247	9268	9290	9311	2	4	6	8	11	13	15	17	19
·97	9333	9354	9376	9397	9419	9441	9462	9484	9506	9528	2	4	7	9	11	13	15	17	20
·98	9550	9572	9594	9616	9638	9661	9683	9705	9727	9750	2	4	7	9	11	13	16	18	20
·99	9772	9795	9817	9840	9863	9886	9908	9931	9954	9977	2	5	7	9	11	14	16	18	20